SpringerBriefs in Materials

AF793807

For further volumes:
http://www.springer.com/series/10111

Mihaela D. Leonida

The Materials and Craft of Early Iconographers

Springer

Mihaela D. Leonida
School of Natural Science
Fairleigh Dickinson University
Teaneck, New Jersey
USA

ISSN 2192-1091 ISSN 2192-1105 (electronic)
ISBN 978-3-319-04827-7 ISBN 978-3-319-04828-4 (eBook)
DOI 10.1007/978-3-319-04828-4
Springer Cham Heidelberg New York Dordrecht London

Library of Congress Control Number: 2014932422

© The Author(s) 2014
This work is subject to copyright. All rights are reserved by the Publisher, whether the whole or part of the material is concerned, specifically the rights of translation, reprinting, reuse of illustrations, recitation, broadcasting, reproduction on microfilms or in any other physical way, and transmission or information storage and retrieval, electronic adaptation, computer software, or by similar or dissimilar methodology now known or hereafter developed. Exempted from this legal reservation are brief excerpts in connection with reviews or scholarly analysis or material supplied specifically for the purpose of being entered and executed on a computer system, for exclusive use by the purchaser of the work. Duplication of this publication or parts thereof is permitted only under the provisions of the Copyright Law of the Publisher's location, in its current version, and permission for use must always be obtained from Springer. Permissions for use may be obtained through RightsLink at the Copyright Clearance Center. Violations are liable to prosecution under the respective Copyright Law.
The use of general descriptive names, registered names, trademarks, service marks, etc. in this publication does not imply, even in the absence of a specific statement, that such names are exempt from the relevant protective laws and regulations and therefore free for general use.
While the advice and information in this book are believed to be true and accurate at the date of publication, neither the authors nor the editors nor the publisher can accept any legal responsibility for any errors or omissions that may be made. The publisher makes no warranty, express or implied, with respect to the material contained herein.

Printed on acid-free paper

Springer is part of Springer Science+Business Media (www.springer.com)

To my father, the trail blazer
Nothing is too small: against a gold background
I paint it large and lovingly
and hold it high, and I will never know
whose soul it may release....
(Rainer Maria Rilke, The Book of Hours, 1905)

Foreword

This little book closes the gap between the heavily documented manuscript and icon painting traditions of Western Europe and of the Middle East. This is evident in its treatment of the preparation and use of painting materials used in the Southeastern European tradition, where it can be seen that its iconographers borrowed heavily from other sources, while at the same time maintaining their own standards, traditional methods and materials. For example, Romanian pigment makers were careful to insist on greater purity of their starting materials, and hence, avoided insofar as was possible, contamination by side reactions in the final product. This practice, in the end, resulted in markedly less degradation of the manuscript or icon over time. Another unique tradition, at least from the 16th century on, was the practice of frescoing churches on their exterior, which demanded not only great care in the preparation of the surface, but also in the selection of the fresco materials—two considerations that even today attest to the iconographers' expertise as evidenced by the excellent condition of these frescoes after over four hundred years of exposure to harsh climatic conditions.

An essential aspect of the book is its description of the cultural background that gave rise to this form of expression, that is, the importance and hallowed character of manuscript and icon creation. Not merely an art form as those from the Western tradition would define it, it was, to its iconographers, a sacred trust in the transmission of faith. Its practice was viewed as a means of catechesis, or teaching the truths of religion, much as the stained glass windows of medieval churches and cathedrals accomplished this function in the West—it was truly communicating a creed to a largely illiterate society through imagery.

Peppered with beautiful images which presumably only someone very close to the Southeastern European painting schools tradition would have access to, this book is a delight to the eye, and at the same time is crammed with primary source information that would be difficult to find in any one work. It is a must-read for anyone interested in everything that went into image creation in antiquity.

New Rochelle, New York — Mary Virginia Orna

Preface

I would like to explain in a few introductory words what made me write this short book and what I am trying to accomplish thereby.

I grew up as the only child of two scientists, a family with modest income, in Romania, a satellite country of the Soviets at that time. All their spare time and resources were invested in their daughter's education in parallel with gathering a modest collection of Romanian artifacts, paintings, icons, peasant decorative objects, drawings. Before I learned to read I could differentiate between icons on glass and those on wood, between the naive realism of paintings representing the working class building the socialist society and those of the abstract-ironical Dadaism from an earlier time in the century. I gradually realized that although the language of visual arts is at least as strong and impressive as the written words, the bridge between the two can only be covered with the help of a guide—a book which explains in easy to understand words the images proposed by visual arts.

The icons always fascinated me. Later on I saw entire churches painted inside and out, I met a few old iconographers and I realized that the role of paintings was even more important in times past than in modern societies. The colored images from churches and monasteries had a major role in conveying ideas about the world, about society, about the moral principles and the culture of many generations before us, many of which included illiterate members as a majority.

Although this is true for all civilizations with European roots, the feelings and thoughts impressed on the viewer by the cathedrals of Western Europe are totally different from those one feels in front of the icons and frescoes from the churches of the eastern part of the continent. The religious fervor of the present day populations from the Russian, Romanian, Balkan or East-Mediterranean areas is probably rooted in the expressiveness of the paintings of the Orthodox East.

This book tries to illustrate—and to explain as much as possible—the material ways used by the Orthodox iconographers of times past (intentionally anonymous, as a rule) who succeeded to reach such impressive heights of the artistic expression, while observing the limits imposed by a very rigorous doctrine.

The hypothesis of the author—a scientist active in the field of methods for matter conversion—is that judiciously choosing the materials (some of them prepared by the artisans themselves) and techniques used in painting was the decisive factor which imparted a dramatic nature and unique expressivity to the icons of Southeastern Europe and helped them stand the test of time.

To the reader, the book opens a window toward the hidden face, the material face, of the Orthodox icons, toward the secretive world of iconographers of times past. I hope that that is interesting and will capture you.

Teaneck, New Jersey, USA Mihaela Leonida

Acknowledgments

Many thanks are due to Mary Virginia Orna for reading the manuscript, for her kind words, and for being an example to follow for all of us scientists working on interdisciplinary topics.

Another strong woman, my mother, was with me through the entire journey, showed me how to be a scientist, a mother, and how not to give up. She still fills my thoughts and my heart.

I would like to acknowledge with great thanks James Yu for his infinite patience, unwavering support, and continuous effort to afford me a better grasp of the intricacies of the English language. His suggestions concerning this manuscript and many others have been valuable and always greatly appreciated.

Last but not least, heartfelt thanks to Maria and Michael for being my home away from home and to Marie-Marie for being my rock.

Contents

List of Figures

Chapter 1
The Technical Part of the Painter's Manuals Compiled and Used by Early Iconographers. Historic Notes

The word "icon" comes from the Greek *eikon* which means image. Typically icons[1] are devotional objects representing characters or scenes from the Bible. Throughout history they have been made on different supports, using different materials and tools.

This book presents the materials prepared and the techniques used by early iconographers of the Orthodox East. Looking at centuries-old paintings and frescoes which stood well the test of time, we marvel at the brilliance of their colors, the quality of their materials, and wonder about what these artisans knew and how they prepared their own materials and tools. What made their works stand the test of time so well (Fig. 1.1)? Coming from a time when few could read, their products, the icons, were an important teaching tool. Very little is known on the subject of the craftsmanship involved in making icons for several reasons. First, those involved in the process were members of a closed circle to which access was achieved only following a long and strenuous apprenticeship. The knowledge was passed from master to apprentice by direct demonstrations and through the information contained in the *hermeneias* (painter's manuals) compiled by iconographers and intended exclusively for their restricted group. Due to the manner and the circumstances in which that knowledge was acquired and, even more importantly, due to the presumed qualities of the products made by iconographers, the icons, the information contained therein was considered a "secret" and could not be shared outside the closed circle of iconographers.

While in Western Europe these "secrets" ceased to be considered as such around the twelfth century, in the Orthodox East the situation stayed unchanged until much later. The iconographers of the East constituted a particular social segment in that area and it was unanimously accepted that only they could create the icon, the sacred art object considered to have a definite functionality.

The Church of the East preached and still preaches through images as well as with words. Icons were considered relays which offered to those who prayed in front of them the unique possibility of establishing a direct connection with the Transcendental [1]. The religious paintings of the Orthodoxy appear also as an access road

[1] Icon—used in the most comprehensive sense of the word, irrespective of the support it was painted on (wood, wall, textiles, etc.), of the technique used (*fresco, secco*, tempera, etc.), irrespective of the place it was exhibited (church, private house, exhibition hall), or of its dimensions.

M. D. Leonida, *The Materials and Craft of Early Iconographers*,
SpringerBriefs in Materials, DOI 10.1007/978-3-319-04828-4_1, © The Author(s) 2014

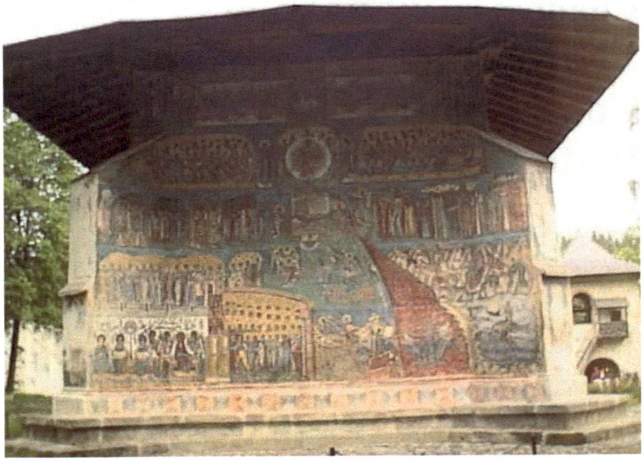

Fig. 1.1 Western external wall of the church at Monastery Voronet (Northeastern Romania), painted in the sixteenth century. Photo: M.D. Leonida

for the masses of believers to the truths of a new religion, a true *Biblia pauperum*[2], a "seen" Bible, of Southeastern Europe.

Access to the iconographers' guild was achieved only after a long apprenticeship concluded by a special consecration (religious) service. During that ceremony a priest (possessor of the *Grace* and of the power of conferring it) bestowed the divine *Grace* (*Grace*) upon the apprentice, brought to the ceremony by his mentor (see Chap. 11). Maintaining the *Grace* was possible through certain psychological exercises and social actions during the entire subsequent activity and life of the iconographer. It was considered that the new recruit to the iconographers' circle, after the consecration ceremony and only after it, possessed the *Grace* which enabled him to create the functional Icon. He could not transmit it to somebody else. For this reason, the iconographers' guild, constituted through different means than those used by other artisans since Antiquity and through the Middle Ages, was strictly limited to the consecrated ones. The knowledge transmitted by masters to their apprentices had been accumulated and conserved through the technical-artistic tradition and handed down many past generations. Each generation enriched the inheritance with information acquired through its own activity and experience as well as with knowledge obtained from peers or from artisans elsewhere. All these contributions constituted a representative artistic entity for a certain geographical area, during a historical period. The whole process was a live one and this fact has been mentioned starting with the preface of the first translated and printed Greek *hermeneia* [2].

Today, long after the painter's manuals were compiled, some of the terms found therein have been forgotten or fallen out of use, which makes some parts of these manuscripts difficult to understand. We are therefore not surprised to find the opinion

[2] Paupers' Bibles—colorful hand-painted illuminated manuscripts on vellum. In the fifteenth century printed examples with woodcuts took over, mostly in the Netherlands and Germany. The simpler versions were used by clergy as a teaching aid for those who could not read, which included most of the population. The name *Biblia pauperum* first used by German scholars in the 1930s.

expressed in some older studies and books, whose authors lacked a technical background, that the recipes recommended by the *hermeneias* are impossible to understand and reproduce [3].

In Western Europe, the first manuscripts with technical information concerning painting appeared early during Antiquity. Their authors were either painters themselves or visitors of their studios and often the information provided was accompanied by technical recommendations. The earliest manuscripts of this type marked the beginning of writing as a means to hand over the knowledge about the painters' art and techniques. This change is a revolutionary one both in the means of communication and in the way "secrets" of painters were considered as such. Writings of Vitruvius[3] and Pliny the Elder[4] are examples of such manuscripts.

Among the manuscripts on this topic which appeared during the early Middle Ages or later outside of Southeastern Europe we have to mention those written by Vasari[5] and Dioscorides[6], the much quoted Lucca Manuscript (eighth century), Heraclius' manuscript[7] (seventh century), and the manuscripts of some Arab and West European alchemists[8]. However, the oldest and most accomplished "book on painting" remains that of the Benedictine monk Theophilus Presbyter (*Schedula diversarium artium*—On Different Arts)[9], written in the eleventh century. From about the same time a treatise on painting attributed to the Chinese painter Kuo Hi together with some six others authored by Chinese painters reached us as well [4].

Obviously the manuscripts we mentioned constituted exceptions for the time they were respectively authored. The general rule of the time was to limit the access to the "secrets" of the trade (painting) to the members of the close-knit circle of painters, the only ones having access to these manuscripts. However, this rule is not spelled out clearly anywhere. How else could we interpret Cennini's[10] dedication of his fifteenth century "Book on Painting" which opens with a clear statement of purpose: "for the use and good profit of anyone who wants to enter this profession" [5].

[3] Marcus Vitruvius Pollio (born c. 80–70 BC, died c. 15 BC), Roman architect, engineer, and author of the treatise *De architectura* (*On Architecture*), a handbook for architects.

[4] Pliny the Elder (Gaius Plinius Secundus)—born AD 23, died AD 79—Roman author, naturalist, philosopher, and army commander of the early Roman Empire. Wrote the encyclopedia *Naturalis Historia* (Natural History) which became a model for encyclopedias.

[5] Giorgio Vasari (1511–1574)—Italian painter, architect, writer and historian. Wrote *Lives of the Most Eminent Painters, Sculptors, and Architects*, important contribution to the field of history of art writing.

[6] Pedanius Dioscorides (c. 40–90 AD)—Greek physician, pharmacologist and botanist. Wrote *De materia medica* (Concerning Medical Matters), a five volumes treatise.

[7] Flavius Heraclius Augustus (AD 575–641). Wrote *De coloribus et artibus Romanorum* (On the Colors and Arts of the Romans).

[8] Alchemy flourished in Greco-Roman Egypt, the Islamic world during the Middle Ages, and then in Europe from the thirteenth to the eighteenth centuries.

[9] Theophilus Presbyter (c. 1070–1125) compiled a detailed descriptions of various medieval crafts, *Schedula diversarium artium*—On Different Arts, providing detailed insights into the techniques used in the applied arts in the Middle Ages.

[10] Cennino d'Andrea Cennini (c. 1370–c. 1440), Italian painter, influenced by Giotto.

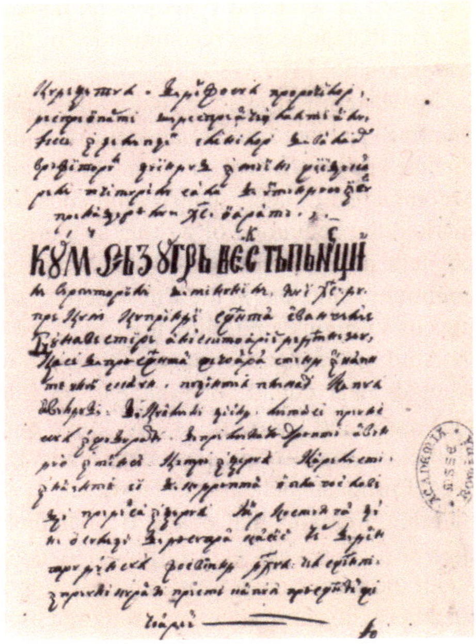

Fig. 1.2 Page from a Romanian manuscript (painter's manual, *hermeneia*). Photo: M.D. Leonida

In the Southeastern Europe area, information and recommendations concerning the activity of iconographers appeared scattered in manuscripts of different types, not only in those dealing with art related topics. The best organized and the most complete such recommendations appear in the *hermeneias*[11], the Romanian and Greek iconographer's manuals. Without knowing too well either when and where they appeared first or when the revolutionary switch from the oral transmission of the knowledge and technical tradition in the religious painting to the written one occured, we only know that they existed and that only a few of them were preserved for us.

It is interesting to note that the only *hermeneias* known are either in Greek or in Romanian (Fig. 1.2), languages of the two southeastern European peoples who were never integrally occupied by the Turks. All contain approximately the same type of material, organized in two big parts (a technical part and a part about iconographic matters).

Here is how the sequence of the discovery of the *hermeneias* and of the many technical details pertaining to the "hidden face" of the Orthodox icon in Western Europe unfolded. In 1832, in a German journal [6], mention was made of an Athonite *hermeneia*. It had been found in the possession of a Greek iconographer who was painting an Orthodox church in Munich. The article did not raise too much interest at the time and remained ignored for about a decade. In 1845 two French researchers published (in French) the translation of a manuscript (*hermeneia*) they

[11] *Hermeneia*, in Greek, means translation, interpretation, explanation. It was used in the ancient Greek-speaking world for detailed, systematic exposition of scriptural content.

had found in the studio of a monk-painter who lived and painted in a monastery, part of the Monastic Republic at Mount Athos [2, 7]. The two French researchers accompanied their translation with numerous and well-documented comments, much more so than Schorn did in his paper announcing the discovery of the first such manuscript by the European West. The French translation of the first Greek *hermeneia* published integrally was dedicated to Victor Hugo, the well-known literary giant of that time. The enthusiasm it generated prompted the French state to cover the publishing costs of that edition.

Starting in 1845, the name *hermeneia* entered the language used by European scholars. It was a name used in the title of such technical-artistic manuscripts in both Greek and Romanian. Under this name they circulated through the entire area of the Orthodox East. After another decade, in 1855, the German translation of the same Greek manuscript was published. The first Greek edition had been published two years earlier, followed, three years later, by a second Greek edition. In 1868 the Russian specialist in paleography and religious art, Porfirii Uspenskii, published the Russian translation of the same manuscript [8]. In the appendices to this edition fragments from two other *hermeneias* (translated into Russian also), elaborated in the seventeenth and the eighteen century respectively, can be found. Also in Russia, this time in St. Petersburg, in 1909 the translation of another Greek *hermeneia* was published. This manuscript has been attributed to Dionysius of Fourna, a monk-iconographer who lived, painted and wrote at Mount Athos [9].

In Russia another type of manuals for iconographers (*podlinniki*) circulated as well. They were much less elaborate than the *hermeneias* and did not include a technical part [10].

The first *hermeneia* printed in Romanian appeared in 1891. It was edited and had a preface by the Bishop Ghenadie of Râmnic. In the preface he was trying to elucidate the origins of the oldest manuals for iconographers of the Orthodox East. By quoting Russian sources, an attempt was made therein to date the discovery of the first Greek *hermeneia* almost two centuries earlier than the discovery of such a Greek manuscript by Didron and Durand. The close connection between the Romanian religious painting and that at Mount Athos was mentioned as well. In 1904, a second edition followed [11] in which this preface was eliminated. It was probably considered obsolete as information. In these two editions, beside information directly of interest to the Romanian iconographers, some technical recommendations of a different nature and some household recipes appear as well.

Between 1882–1932, in the Romanian area, studies were published referring to either "painter handbooks" or to fragments of *hermeneias*. They were published by Kogălniceanu [12], Ştefan Berechet [13], Iorga [14], Petrescu [15], and others. Later on, the same Costin Petrescu, as well as Sandulescu-Verna [9], Voinescu [16], and others published more studies concerning the activity of the Romanian iconographers in general and their manual of painting, taking particular interest in their iconographic part.

In 1984 Mihalcu published a book [17] dealing in more detail with the technical "secrets" of the Romanian medieval iconographers. Twelve years later, in an encyclopedia [18], the same author explained archaic technical terms used in the Romanian *hermeneias* and the workshops of iconographers from times past. The publication of this book shattered the theory that the *hermeneias* cannot be understood anymore today because the meaning of some technical archaic terms, which are no longer used, has been lost or because the facts presented therein are not supported by present day science.

Today we can say with certainty that the *hermeneias* are compilations including important contributions from their elaborators. These manuals contain information and recommendations originating from their own personal experience together with those acquired from colleagues and predecessors. Thus, in a Romanian manuscript a *poliment* was "handed down by old masters" and in another one [19] the elaborator says that he learned the recipe for an adhesive used in gilding "while working with a Russian iconographer". Very seldom is reference made in a recommendation to the name of an earlier religious painter. One of these rare cases is that of the standard color "sarca", a material called "Panselinos'[12] sarca" in a Romanian *hermeneia* [20].

In the rough conditions in which they were utilized (painter studios or building sites of churches), the *miscellanea* (different types of manuscripts containing varied information) and the *hermeneias* wore out easily. Sometimes it happened that the information and the recommendations contained therein disappeared and were forgotten. Still other times they were transcribed by the temporary owner of the manuscript or by somebody else. On these occasions, some information considered not interesting or obsolete was eliminated and results from the experience of the transcriber or gathered by him from other sources (local workshops or from other parts of the country or Europe) were added.

The rapport between the elaborator of the manuscript and its reader was the same as that existing in the case when communication and instruction were conducted live and by practical example, from a master to his apprentice. This is why the same obligations arose. The master was morally obliged to give to the apprentice (to any reader) everything he had learned from his own mentor, from other iconographers, and from his own experience. This is very clear from the forward (*predoslovie*) of some *hermeneias* [21], and from the writings of the iconographer Petrescu [22], one of the very few possessors of the "secrets" of the old iconographers, who put them on paper in more recent times.

When carefully analyzing the differences—greater or smaller—between manuals for iconographers from the European Southeast, the researcher is often tempted to consider these manuscripts, called translations by their elaborators, as compilations. This impression stems from some additions or explanations which obviously are the result of the author's own activity, in spite of his humbly calling himself a mere "translator". Thereby, some of these declarations can be considered, especially those from the technical part of the *hermeneias*, as required in order to increase the audience and as a guarantee to the seriousness of a particular manuscript.

[12] Manuel Panselinos—Athonite painter from the Paleolog's epoch (thirteenth century). He was known as one of the greatest painters of frescoes. Most of these can be found in the Protaton Church, at Mount Athos.

References

1. Mihalcu M, Leonida MD (2013) The hidden face of form and color. Eikon, Cluj-Napoca, p 158
2. Didron P, Durand AN (1845) Manuel d'iconographie chretienne grecque et latine. Imprimerie Royale, Paris
3. Berger E (1987) Quellen und Technik der Freskos, Oele, und Tempera Malerei des Mittelalters nach Quellenschriften und Versuchen. Reprint 2009 of the 1912 edition, Medieval and Modern Times Series. Saendig Reprint Verlag, München
4. Grunier J (1970) L'art et ses problemes. Rencontre, Paris
5. Cennini CA (1960) The craftsman's handbook. Translated by Daniel Thompson. Dover Publ. Inc., New York
6. Schorn L (1838) Kunstblatt, Stuttgart und Tübingen 1–5(2):431
7. Grecu V (1936) Carti de picture bisericeasca bizantina. Glasul Bucovinei, Cernauti
8. Uspenskii PD (1868) Erminiia ili nastavlenie v jvopinomi iscutssvo sostavlenoe ieromonahomi i jivpistemi Dionisiemi firnoagrafiotom. Kiev
9. Sandulescu-Verna C (1979) Erminia picturii bizantine. Mitropoliei Banatului, Timisoara
10. http://russianicons.wordpress.com/2011/09/29/podlinniki-the-manuals-of-icon-painting-and-how-to-read-them/. Accessed 19 July 2013
11. Ghenadie of Râmnic (1904) Iconografia. Arta de a zugrăvi biserici şi icone bisericeşti. Tipografia cărţilor bisericeşti. Bucharest
12. Kogălniceanu M (1882) Revista pentru istorie, arheologie si filozofie 1:33–35
13. Berechet S (1928) Buletinul Comisiunei Monumentelor Istorice (Basarabia), II: 125–135
14. Iorga N (1921) Comisia franco-romana II, nr. 33
15. Petrescu C (1932) Universul (Bucuresti), XLIX, May 1, 20: 9
16. Voinescu T (1978) Radu Zugravu. Meridiane, Bucharest
17. Mihalcu M (1984) Romanian medieval values. Sport-Tourism, Bucharest
18. Mihalcu M, Leonida MD (2013) The hidden face of form and color, 2nd edn. Eikon, Cluj-Napoca
19. Romanian manuscript 1127 The Library of the Romanian Academy, Bucharest
20. Romanian manuscript 3586 The Library of the Romanian Academy, Bucharest
21. Romanian manuscript 4353 The Library of the Romanian Academy, Bucharest
22. Petrescu C (1931) L'art de la fresque. Lefranc, Paris

Chapter 2
Materials From Natural Sources and Those Prepared in the Iconographer's Studio or in the Peasant Household

The materials used by ancient iconographers in the Southeuropean area and which are described in the painter's manuals and in *miscellanea* were found in the surrounding nature (ores, local flora) or in the peasant household (beeswax, linseed oil, tallow, soap, vinegar, ethylic solutions). Some materials (pigments, resins, some solvents, siccatives, fillers, dehydrating agents, flocculants) were prepared in their workshops or in those of other iconographers, local or foreign.

The same types of sources for raw materials as those mentioned in the *hermeneias* were used by painters in other parts of Europe during Antiquity and Middle Ages, not only by the iconographers in Byzantium, as stated in most publications. An example is that of cinnabar which was extracted from ores, as mentioned in the writings of Pliny and Theophrastus[1] and in the *hermeneias.* The preparation of synthetic cinnabar (see Sect. 3.1) as described in the Romanian painter's manuals is very similar to the process used by the Chinese beginning in the first centuries of the last millennium [1]. The same was true in the case of white lead[2] (called "icon white" by iconographers). When it was prepared in the painter's workshop, the technology was the same as that practiced in European and Chinese Antiquity [2]. Even the names used in different areas of Europe were similar (*ceruza*—in the Romanian area, *Venetian white, Nottingham white, Vienna white, Cremnitz/Kremnitz white, ceruza*—in Hungarian, *cerussa*—in Latin, hydrocerussite—in modern times, for the complex mineral). At the same time, the way this process is described in different *hermeneias* has the interesting and original ring of an old technical tradition specific for a certain area and bearing the preferences of each artisan involved in the compilation of the technical manuscripts.

The differences in names can appear in connection with different commercial sources for the materials used by iconographers. This situation was influenced by the ever-changing socio-political conditions in Southeastern Europe. Differences in materials acquired from commercial sources would result in variable compositions and technical characteristic of the compound of interest, depending on the source available at a certain time.

[1] Theophrastus of Eresos (lived around 300 AD) and is known especially for his treatise *De Lapidibus* (History of Stones), the earliest book on minerals.

[2] White or light grey mineral, occurring naturally, with the chemical formula $2PbCO_3 \cdot Pb(OH)_2$.

M. D. Leonida, *The Materials and Craft of Early Iconographers*,
SpringerBriefs in Materials, DOI 10.1007/978-3-319-04828-4_2, © The Author(s) 2014

In the technical part of *hermeneias* materials are characterized sparingly, unlike what is offered in similar technical publications at the present. In such a manuscript [3], for painting on mother-of-pearl, it was recommended to prime the surface using a "watery egg (an aqueous solution of egg) for the paint to be accepted". At other times, the elaborators of painter's manuals indicated only one or two characteristics of a material, whatever was easy to unequivocally evaluate by one of the five senses. A glue used in wood joining needed to be "strong" (high concentration) [4] while an ochre used for hair and beards needed to be "burned" (converted by roasting into a darker red Fe_2O_3) [3], and a lime was supposed to be not slaked (not hydrated) [5].

Sometimes a raw material was defined by indicating the town, country, or geographical area it was obtained from. That was the case when the material was from natural sources, or when it had been prepared in another workshop, either domestic or foreign. An ochre needed for the recipe of a standard color *Panselinos' sarca* was *Venice ochre* [3], a second ochre, part of another standard color (*proplasma*) was known as *ochre from St. Anna* [5], while yet another ochre needed for an adhesive was *Tarigrad* (Constantinople) *ochre*.

The iconographers were very knowledgeable and their guidelines did not leave room for any misunderstanding. In a Romanian technical manuscript [6] two different recipe components needed to be obtained from Constantinople (today's Istanbul), a "white earth" (chalk), needed to prepare a *poliment,* is said to be from "Ipu's slum" (a neighborhood in Constantinople). This is meant to differentiate it from *Tarigrad ochre*, another material imported from Constantinople and mentioned in the same manuscript. Yet another ochre (*Thassos ochre*) was different from the *Tarigrad ochre* due to a darker red color (higher content of Fe_2O_3). Use of such a name shows, beside the commercial connections in this part of the world at the time the handbooks were elaborated, the same deep knowledge of the materials and their properties. When *Venice turpentine* is needed in a varnish recipe, it is indicated that it would result in an enamel-like look of the painted layers (due to high acidity index of the raw material).

For easily accessible materials, *hermeneias'* elaborators did not bother to specify from where to obtain them. Limestone (natural calcium carbonate, $CaCO_3$) necessary for many primers, ox bile necessary for some adhesives used in gilding, brandy and spirit needed for varnishes, all were considered well known and easy to find. This attitude is common to authors of painting manuals in the entire medieval world. Cennino Cennini in his Handbook [7] does not specify from where limestone or gypsum (calcium sulfate) can be obtained in the same manner as Dionysius of Fourna[3] [8] and Theophylus Presbyter [9].

Conversely, for some materials the manuals give a trove of information, necessary and a proof that the anonymous elaborator had a remarkable capacity to observe and a matching technical ability, drawing on the experience of several generations of artisans. An example can be found in a manuscript [10] which describes, among other processes, how to make a glue necessary to make grounds,

[3] Dionysius of Fourna (*c.*1670–c.1744) was an Eastern Orthodox author of a *hermeneia* (compiled at Mt. Athos 1730–1734) and painter of icons and churches.

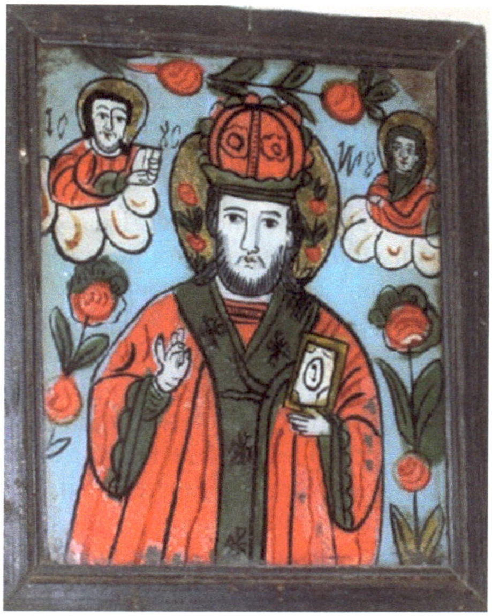

Fig. 2.1 Romanian icon on glass (beginning of nineteenth century) showing garments and flowers painted using cinnabar containing paints. Photo: M.D. Leonida

including the conditions required, and the fact that it has to be prepared immediately before use (otherwise it destabilizes the ground). The pieces of leather used to extract collagen needed to be "well tanned", preferably ox leather ("the best"), and the water used for the extraction had to be "clean" (without suspensions). A sketchy, comparative discussion of pieces of leather from different sources can also be found therein, complete with economical considerations.

Concerning gypsum from natural sources, a manuscript [3] recommended choosing white and shiny pieces (with no colored impurities and well crystallized), a very pure material which the centuries-long tradition considered to be the most desirable for primer preparations. Similarly, a *bole*[4] needed to be shiny and, when used as a filler, it had to be "not sandy" (low content of silicates) and "not too red" (low content of Fe_2O_3) [3]. The latter was a requirement also when it was used to warm up the color of the gold leaf, in gilding (see Sect. 4.2).

For the cinnabar (native mercury sulfide HgS) used by Romanian iconographers of the past to paint in red garments (Fig. 2.1) of different characters, among other names, the painter's handbooks use "tzinober stone" which probably refers to German or Austrian sources (*Zinnober* in German) and what the popular etymology made of the borrowed name. When describing this inorganic compound, the elaborator of a *hermeneia* mentions large crystals with a bluish tint, easy to evaluate by sight and characteristic for a pure HgS [10]. The dangers due to adulteration with minium (Pb_3O_4), red-orange as well but cheaper and prone to darkening, are also discussed therein (see Sect. 3.1).

[4] A red earth composed of iron oxide and clay, used in gilding (in water) as a ground. Also known as *Armenian bole* and *poliment*.

Other materials found in natural deposits beside ochre, chalk, and cinnabar were zinc sulfate ($ZnSO_4$—used as a siccative for linseed oil), reactants used in chemical syntheses conducted by the artisans in their workshops (mercury, sulfur, copper, lead), and iron sulfate (*copperas*[5]) used in the recipes of different inks.

Materials used as solvents (concentrated ethylic solutions such as brandy), concentrated acetic acid solutions used as disinfectants, "strong vinegar", beeswax, tallow, soap, bovine bile, eggs, linseed oil for varnishes, fillers such as straws and oakum used in mortars, horse dung (providing heat and carbon dioxide during its fermentation), all these were obtained from their own peasant household. The technical part of the handbooks offers detailed characterizations of the materials obtained by different unitary operations in the household. For example the vinegar used in the preparation of lead white ("icon white") has to be "strong" (high acetic acid concentration) [3]. This is proof that the iconographers' experience recorded the direct dependence of the reaction kinetics on the concentration of the reactant. For spirits, the concentration required for a certain application is sometimes stated qualitatively as "very strong brandy" for a varnish [10] or more specifically by indicating the number of distillations required for the alcohol to reach the desired concentration in the aqueous solution. All these recommendations are offered in a moving, archaic language which, on thorough examination, stands the test of time and makes the statements easily translatable into the language of present day science. In a *"hermeneia"* [3] it is mentioned that a "strong brandy" was obtained "from another brandy (double distillation)" and an alcoholic varnish (see Chap. 6) required "very strong brandy ..... which was turned out four times" (obtained by four successive distillations). For a slaked lime recommended for a mortar, the consistency (viscosity) had to be that of a clay suspension and, when tasted, "had to be neither bitter, nor tart" [11].

Among the materials prepared using chemical processes in iconographers' workshops we mention white pigments, different varieties of lime white [12], lead white (basic lead carbonate), synthetic cinnabar (*vermillion*), copper acetate (*verdigris*, "lawn green", "brass green"), and slaked lime.

Among the materials collected from nature were: flax seeds used to extract linseed oils (for paints and varnishes, see Chap. 6), balsam (resin collected by tapping pine trees) which was used to obtain turpentine[6] by distillation, and colophony. When the materials were harvested from organs of local flora, the *hermeneias* were indicating the period during which they had to be obtained for optimal performance. A Romanian manuscript [10] recommends to use garlic cloves harvested in June-July to allow them to develop and reach the highest content of the protein needed when their juice was used as adhesive in gilding (see Sect. 4.4). A Greek *hermeneia* [8]

---

5 Ferrous sulfate ($FeSO_4$), known since ancient times as copperas and as green vitriol (due to the blue-green color of its heptahydrate). Upon heating, loses the crystallization water and becomes a dark brown solid.

6 Colorless solvent obtained by distillation of pine resin in ancient Greece and Rome; it was obtained by boiling pine pitch covered by a sheepskin, fleece side down. The liquid condensed on the fleece was harvested by wringing the sheepskin periodically.

recommended to do the harvesting in June which implies the same calendar guidance in both cases since Greece lies at a more southern latitude than Romania.

Beside the imported materials already discussed, revealing in their names their place of origin, other materials and raw materials were imported as well. But always, the only materials imported were those which had no autochthonous source, such as specific pigments (*vermillion*, some ochres, *ceruse*), exotic resins (dammar, sandarac, aloe, shellac), spermaceti, or indigo.

In general, the technical part of the painter's manuals displays a very diverse list of materials and raw materials, compiled after a centuries-long practice of maintaining connections with peers from other parts of the Orthodox East, which was handed down from generation to generation within the closed circle of iconographers.

References

1. Eastlake CL (1967) Materials for a history of oil painting. Dover Publ., New York
2. Holley CD (1909) The lead and zinc pigments. First ed., Wiley & Sons, New York, p 2
3. Romanian manuscript 1795 Library of the Romanian Academy, Bucharest
4. Romanian manuscript 4206 Library of the Romanian Academy, Bucharest
5. Romanian manuscript 5769 Library of the Romanian Academy, Bucharest
6. Romanian manuscript 2151 Library of the Romanian Academy, Bucharest
7. Cennini CA (1960) The craftsman's handbook. Translated by Daniel Thompson. Dover Publ. Inc., New York
8. Dionysius of Fourna (1974) The painter's manual. Translated by Paul Hetherington. St. Vladimir Seminary Press, Yonkers
9. Thompson DV (1967) Theophilus Presbyter, words and meanings in technical translation. Speculum 42(2):313–339
10. Romanian manuscript 1555 State Archives, Bucharest
11. Romanian manuscript 1808 Library of the Romanian Academy, Bucharest
12. Mihalcu M, Leonida MD (1980) Lime and technologies connected to its preparation and uses in the recommendations of the *hermeneias* and in the Romanian medieval painting. Rev. Muzeelor si Monumentelor Istorice si de Arta, XLIX, 1, 61

Chapter 3
Pigments

The first use of color precedes written history and it probably involved body painting. Early humans found their pigments in the minerals occurring in soils and clays. Starting with the Cro-Magnon and Neanderthal humans, red ochre, hematite, Fe_2O_3 [1], appears in burials or fertility rites, probably representing blood. Continuing along these lines, the next step was the early use of color in the cave paintings of Chauvet and Lascaux in France, Altamira in Spain, and those at Tassili n'Ajjer (North Africa). When humans learned how to grind minerals, almost thirty thousand years ago, they mixed them with water (the first pigment vehicle) and applied them to stone surfaces [2]. These earth colors were the earliest known coloring materials and they lacked brightness. Greens, blues and vivid colors in general were totally absent from the first palettes used by our ancestors. The missing colors were discovered by the civilizations around the Mediterranean [3].

In general, very little attention has been given to how artists and artisans obtained their colors compared to how they used them. In the technical part of the *hermeneias* detailed procedures for obtaining pigments are described in an old and magical language. They stood the test of time and, centuries later, their recommendations can easily be translated into the language of modern science.

3.1 Cinnabar

Archaeological findings showed that cinnabar, mercury sulfide (HgS) from natural sources has been used in the European area since Neolithic times (in Turkey 6300–5500 BC [4], in Serbia 4800–3500 BC [5]). The main source of natural cinnabar in the ancient world was Spain [6]. Written documents state that, during the last part of the first millennium, in China, both natural and synthetic cinnabar were used. The same sources, and the first scientific treatise about minerals authored by Theophrastus (see Chap. 2), describe how to obtain quicksilver (mercury) from natural cinnabar and synthetic cinnabar from elements [7]. About the naturally occurring cinnabar, the same author says that it was obtained from difficult to access cliffs by

M. D. Leonida, *The Materials and Craft of Early Iconographers*,
SpringerBriefs in Materials, DOI 10.1007/978-3-319-04828-4_3, © The Author(s) 2014

shooting arrows to dislodge it [8]. Manuscripts of Arab alchemists like Zosimus[1] and Geber[2] or those attributed to Democritus speak in detail not only about the synthesis of cinnabar from elements but also about the fact that this mercury sulfide has two allotropic forms and that the red form can be obtained from the black one by sublimation [9]. In Western Europe written information about the synthesis of cinnabar appeared in well-known manuscripts such as *Compositiones ad Tingenda musiva pelles et alia*[3] and *Mappae Clavicula*[4] and also arrived through the Arab connection. All the rage of alchemy—with its miraculous elixirs, with the equally sensational "philosopher's stone" and with the numerous poetical statements connected to the technique and the medicine of the time—covered the large spaces of a world networked by complex material and spiritual connections. This is the frame within which the knowledge about cinnabar was transmitted.

At a time when science was still in its infancy, certain virtues were attributed to cinnabar, virtues in which we do not believe anymore. Once in Europe, the knowledge about this pigment spread very fast. A few centuries later, Cenninni's treatise as well as other contemporary manuscripts originating in different parts of Europe considered that synthesis of cinnabar was so well-known that they did not even take the trouble to describe it [10].

Cinnabar was known and used much by the iconographers of the Middle Ages and by those who followed them. They used it in wonderful and resistant frescoes [11], in easel painting and in manuscripts. Whenever they wanted to suggest sumptuousness, the surfaces covered with cinnabar—or with colors having cinnabar in their composition—were large and accompanied by the precious yellow of gold leaf (Fig. 3.1) [12]. They knew its properties, its sensitivity to solar radiation, its chemical incompatibilities, as well as the "dry" process used at the time to obtain it.

Although it is not very clear how the information about the synthesis of cinnabar came to Mount Athos, it is very likely that in the rest of Orthodox Europe it arrived from Athos, on the pages of *hermeneias* or through the experience of painters trained there, and was strongly influenced by Byzantium. Supporting this theory is the name *chinovar* used in the *hermeneias* for cinnabar, a word of Greek origin [13]. Other names (tzinobru, tzinober, tzinober stone) originate in the German name for cinnabar and bear the stamp of the popular etymology.

Sometimes the name *chinovar* was used for the fine cinnabar powder, synthesized and ready to be used to prepare paints. The other three names were often used to refer to a pigment not ground yet and which did not contain minium or other pigments.

---

[1] Zosimus was an early alchemist who lived around 300 AD in Panopolis, Egypt, and wrote the oldest books known on alchemy.

[2] Abu Musa Jabir Ibn Hayyan (Geber), Arab alchemist who lived around 721–815 AD on the territory of today's Iraq.

[3] Collection of pamphlets on ancient occupations and technologies (eighth century AD).

[4] Medieval compilation containing, among others, recommendations for painters (seventeenth century AD).

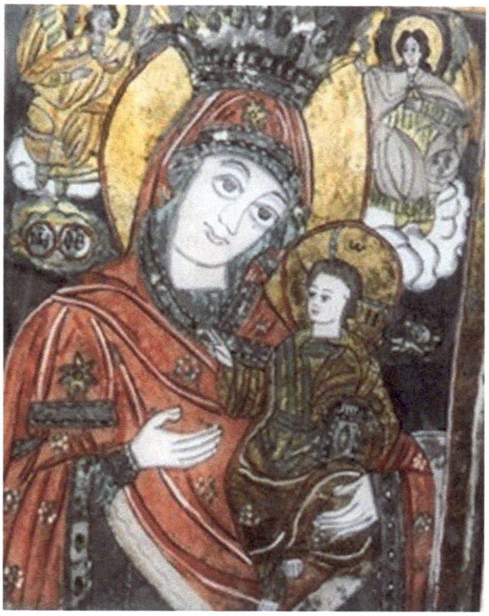

Fig. 3.1 Cinnabar used together with gold leaf, in garments, to suggest sumptuousness (Romanian icon on glass, middle nineteenth century). Photo: M.D. Leonida

Differently from the writers of Antiquity, in the Romanian *hermeneias* there is never confusion between cinnabar, "dragon's blood"[5], and minium[6], all of them red pigments. The difference between cinnabar and minium is emphasized by calling the latter "lead minium". This distinction shows both knowledge of the composition of minium as well as the authors' effort to avoid confusions like those present in other European early painter's manuals [14].

When cinnabar was mixed with lime white, the mixture was called "cinnabar minium". The name describes the result of the mixing: lightening of the color which consequently approaches that of minium. Even more precise details were offered in the painter's manuals about the composition as well as the advantages and disadvantages of using this pigment mixed with others in different techniques of painting (mural, interior or exterior, easel, etc).

Some observations are in order concerning the location of the technical recommendations within these handbooks of painting and technique. The *hermeneias* which appear to be inspired to a certain extent by some versions of the famous book by the Athonite writer Dionysius of Fourna [13] present separately the recommendations concerning the synthesis of cinnabar [15, 16, 17]. One of them mentions using cinnabar in the recipe of a red ink [16]. Other manuscripts provide a sketchy technology for the synthesis of cinnabar when they present the preparation of red inks (see Chap. 7). Among these are included *miscellanea* which look like collections

[5] A bright red resin that is obtained from different plant genera: *Croton*, *Dracaena*, *Daemonorops* (most used), *Calamus rotang* and *Pterocarpus*. It was used in ancient times as dye, varnish, and wound-healing agent. Confused with *cinnabar* in Ancient Rome.

[6] A bright red/orange lead (II, IV) oxide (Pb_3O_4, or $2\ PbO \cdot PbO_2$), also known as *red lead*.

of recipes [18]. A third category of texts does not give any information concerning the synthesis of cinnabar [19]. In the manuscripts strongly influenced by Dionysius of Fourna the recommendations concerning the synthesis of cinnabar are located at the beginning of the "book", together with the other technical recommendations; in other Romanian *hermeneias* they are located at the end [20]. The same situation is true for the *hermeneias* compiled and used by the school at Mount Athos. These differences may exist because, at a certain time, the necessity appeared for cinnabar to be prepared by the painter. There could have been a moment when the pigment either was not available from commercial sources or was adulterated, with or without fraudulent intention.

The synthesis of cinnabar from elements, found in the Romanian *hermeneias*, seems simple enough. Sulfur and mercury were heated together. A final sublimation was converting the black mercury sulfide (β allotropic form, adopts zinc blende structure) into the red mercury sulfide (α allotropic form, hexagonal, optically active).

Three ways to conduct the synthesis have been described.

Variant 1 is the most frequently mentioned and used. It is similar to those used by the Chinese, Arabs, Flemish, and English. The differences will be described shortly.

Variant 2 recommends a small modification in one of the first technological steps (the heating of the reagents is made after their mixing).

Variant 3 mentions the use of a third substance, beside the actual reactants. It serves a purpose which will be discussed later.

In all the variants the core is the same "dry" method known for centuries, in the whole world, and transmitted from one generation to another. At the same time, in different sources, small modifications appear which respond to local requirements or are due to the innovations made by some artists who were continuously looking to improve their materials and the processes used to produce them. The proportions recommended for the mixture of reactants are not the same in different Romanian *hermeneias* and *miscellanea*. This situation is similar for different painter's manuals or manuscripts of alchemists from all times and places. Table 3.1 presents the ratios given by *hermeneias, miscellanea*, or other similar texts of different origins. Pronounced similarities but also differences are noted and it is difficult to decipher the origins or the effects thereof.

Comparing recipes from West European sources with those from the Arab Orient, those from Mount Athos, or those from the Romanian Principalities, it becomes clear that the proportion for a reactant could vary from simple to double (Table 3.1). While in the Romanian, Athonite or Arab manuscripts the differences are sizable, in the Flemish sources they are less different, this showing an apparent standardization of formulations in that region.

In all the sources cited, that is in different places at different times, with the exception of some Russian and Flemish sources, the ratio of sulfur to mercury is much higher than the stoichiometric one. An explanation of this fact can be that, in a process where difficult to control heating was required, high losses of sulfur from oxidation and vaporization may occur. As we shall see shortly, the temperatures required for this type of losses to occur were indeed reached.

Table 3.1 Reactants ratio for the "dry" synthesis of cinnabar, in different manuscripts

Source		Recommended ratio (%)				Destination of cinabar	Observations
Type	Reference	Sulfur	Mercury	Litharge	Sanguine		
Romanian *hermeneias* and *miscellanea*	18	14.80	75.20	6.00	–	Pigment	Ion Halepiu
	20	50.00	50.00	–	–	Ink	
	18	50.00	50.00	–	–	Ink	
	15	14.80	75.20	6.00	–	Pigment	
	17	14.80	75.20	6.00	–	Pigment	Macarie from Căldăruşani
	22	14.80	75.20	6.00	–	Pigment	
	21	50.00	50.00	–	–	Ink	Ghena die of Râmnicului and New Severin
	15	14.80	75.20	6.00	–	Pigment	
Greek *Hermeneias*	13	14.80	75.20	–	6.00	Pigment	Dionysius of Fourna
	23	15.62	78.12	6.26	–	Pigment	
Dutch painter's manuals or alchemy manuscripts	24	16.66	83.34	–	–	Pigment	
	25	14.76	85.24	–	–	Pigment	
	26	12.50	87.50	–	–	Pigment	
English painter's manuals or alchemy manuscripts	27	15.79	84.21	–	–	Pigment	Chamber's Cyclopedia 1728
	27	42.85	57.15	–	–	Pigment	
Manuscripts of the Arab alchemists	29	66.00	34.00	–	–	Pigment	
	30	50.00	50.00	–	–	Pigment	
	31	88.89	11.11	–	–	Pigment	
	32	50.00	50.00	–	–	Pigment	
Russian *podlinniki*	28	14.89	85.11	–	–	Pigment	

The Flemish, using better ovens which allowed some control of the heating, were able to limit the losses of sulfur and to maintain a temperature not much above the optimal one. This was not the case with the Russian artisans. Still, no definite rationale concerning the reasons quantities of sulfur closer to the stoichiometric ones, and lower than those used by other schools, were given in either of the cases. At the same time there are even Flemish recipes asking for more sulfur to be introduced [7, 24, 27, 33, 34].

In the Arab manuscripts the situation is even less clear. The last "book" attributed to Democritus recommends proportions which are exactly opposite to those we consider today as necessary [35]. At the same time in other manuscripts attributed to the same author, as well as in a Syrian treatise on alchemy elaborated about the same time, ratios similar to those recommended by Dionysius of Fourna and by two Romanian *hermeneias* can be found [15, 16, 30, 32].

Other Arab sources [29] either do not specify the proportion of sulfur or, as in some Romanian manuscripts [18,20], indicate a mixture in equal parts of sulfur and mercury. Sources most influenced by Dionysius of Fourna's *hermeneia* recommend the lowest ratio of sulfur, similar to the western and the Russian manuscripts. However, none of them recommends introducing a third reactant, litharge[7], as did the first *hermeneia* published in the West [23]. Concerning this matter a clarification is required: litharge is not a reactant in the synthesis of cinnabar.

It is common knowledge that, in both the Western and Eastern Europe, during the Middle Ages and later, cinnabar was much more expensive than minium (Pb_3O_4). In the Romanian Principalities, even as late as the 19th century, it was three times more expensive. Since both were red pigments, to decrease the expense for red colors, some minium was mixed with cinnabar. This operation was done sometimes with fraudulent design, especially since minium has a higher density than cinnabar (9.10 g/cm^3 vs. 8.10 g/cm^3).

During the synthesis of cinnabar, due to heating, litharge was converted into minium following the equations (3.1) and (3.2):

$$2PbO + O_2 \xrightarrow{400\ ^\circ C} 2PbO_2 + 58.74\ kJ/mol^8 \qquad (3.1)$$

$$PbO_2 + 2PbO \longrightarrow Pb_3O_4 \qquad (3.2)$$

The 400 °C temperature was definitely attained since the following allotropic conversion took place (3.3):

$$\beta HgS \xrightarrow{380\ ^\circ C} \alpha\ HgS \qquad (3.3)$$

The temperature necessary to obtain minium, higher only by 10 % compared to that required to synthesize cinnabar, was attained due to the rudimentary conditions in which the heating was conducted. It seems certain that the purpose of introducing litharge—where recommended—was to adulterate and cheapen the cinnabar. The centuries old experience seems to have revealed yet another proof of the remarkable technical ability whose theoretical bases could only be known several centuries later.

Some researchers wrote that, in England, intentional adulteration of cinnabar with minium started during the seventeenth century. In the Balkans, as in Cennini's Italy, the procedure appeared earlier. This is why it is difficult to decide the reason paints in which cinnabar was used darkened over the centuries. Chemical and mineralogical analyses can indicate if:

[7] Litharge is one of the natural mineral forms (tetragonal) of the dimorphous lead (II) oxide, PbO. The other allotropic form is massicot (orthorhombic).

- the darkening is due to the influence of solar light on pure cinnabar;
- the darkening is due only to the formation of black lead sulfide (in the case of a cinnabar mixed with minium or another lead-containing pigment) or another black sulfide (e.g. copper sulfide);
- the darkening is due to both causes.

Syrian, Russian and West European manuscripts do not contain the recommendation to introduce litharge during cinnabar synthesis but the first Greek *hermeneia* published in Western Europe during the first half of the nineteenth century does.

Sanguine[8] was sometimes introduced during the synthesis, also to reduce the cost of cinnabar [13].

The vessels (in which the stepwise heating during the synthesis of cinnabar took place) were made of burnt clay [13, 15, 17, 36]. In newer manuscripts sometimes an "enameled pot" is mentioned [20, 21], or even a "small jar" [16] which makes the reference to a ceramic vessel even more obvious. The Arab alchemists and the Athonite iconographers used the same material [13, 23, 29, 30]. Only in one Arab text marble is mentioned with the recommendation to apply a layer of clay on top of it [31]. This makes burnt clay the material which actually came into contact with the reactants and the cinnabar. How adherent it was and how much of it was coming off and adulterating the cinnabar is not clear today.

The English "books" recommended the synthesis to take place in iron vessels and the allotropic change in glass containers.

For the reaction vessel, the English alchemists recommended a large opening [37] as did two Romanian *hermeneias* [16, 36], little concern being shown for losses of materials. Conversely, other texts were recommending the vessel to have a small opening and a long neck, to allow for a more efficient cooling [11, 17, 38].

The fuel used for the reaction was either common oak (Quercus) or a variety thereof Quercus frainetto [15, 38], both wood essences with low water content. A similar recommendation appears in some Greek *hermeneias* [13, 23] but does not in either English texts (handbooks of painting or alchemy), nor in the Flemish or Arab ones [26, 37, 38].

The heating was done in improvised ovens, differently from what the Middle Eastern and the Western European alchemists were using. In the latter case the heating was done in ovens specially built for this purpose, where higher temperatures could be reached, and a better control and lower losses were possible. From this point of view, again, the Romanian *hermeneias* and the Russian *podlinnikis* are similar to the Athonite texts.

The placing of the reaction vessels in the middle of the heat source is briefly but clearly indicated by the *hermeneias*. It is stated that the container has to be surrounded by the heat source, up to the height reached inside by the reactants [13, 16, 17). The upper part becomes a cooling space and also the space where, once formed, the pigment is deposited.

---

[8] Sanguine—red colored mineral, found in the caveman's palette and used until seventeenth century as dye stuff. Leonardo da Vinci was the first to use it as a drawing tool.

Concerning the preparation of the reactants prior to the synthesis the recommendations are sketchier. It is only indicated that the solid reactant has to be brought to a "dust"-like granulation [13, 16, 17].

Several manuscripts require the reactants to be heated separately before being mixed [13, 16, 17, 35]. One *hermeneia* recommends to start the heating only after mixing the reactants [18]. On this matter, the description is similar to that found in the manuscripts of the Arab alchemists and Athonite painters [29, 30].

The heating of the reactants was done until the sulfur melted (112.8 °C) [1]. This allowed intimate mixing, with easy to imagine consequences, and minimal losses of sulfur and mercury through oxidation and evaporation [16, 22, 36]. In the cases when litharge was introduced too, the *hermeneias* are not very clear. It is hard to believe that its melting was sought after and accomplished (888 °C), especially if we take into account the way the heating was conducted.

The particular care shown for the intimate contact between reactants and, consequently, for the economy and efficiency of the process, is apparent when familiar terms of comparison like porridge or fine and homogeneous mortar are used to describe how the homogenized reactants were supposed to look [15, 16, 17]. The same care is shown when mention is made that at a point, after a first heating and mixing were completed, the reactant mixture was drained, cooled and ground on a smooth slab of stone.

During the heating the reaction mixture was stirred, in an open vessel, with an iron rod as in the Arab, Flemish and Athonite manuscripts, and differently from the English procedure in which a small wooden shovel was recommended. It must be noted that higher temperatures were attained in the Romanian technology compared to the English one, in which the value for wood charring was the limit.

The mixing was continued until the iron rod became black, just as described in the Dionysius of Fourna text [15, 17, 36]. With the specification of this strange but sufficiently accurate method to control the duration of the heating it seems that yet another way to control the temperature of the heating is offered. The "blackening" of the rod can only refer to the appearance of a black iron oxide (Fe_3O_4, magnetite) following the reaction (4):

$$3Fe + 2O_2 \longrightarrow Fe_3O_4 + 1117\ \text{kJ/mol} \tag{3.4}$$

In the conditions existing in the reaction vessel iron sulfide, which is black too, could be formed as well. It burns in the 400–500 °C interval, producing sulfur dioxide and reddish ferric oxide (Fe_2O_3). Since none of the *hermeneias* mentions the appearance of a reddish compound, we believe that the heating did not go beyond 400 °C. Furthermore, it probably was below 400 °C. None of the old texts of the Arab alchemists mentions this original method of control for the temperature and duration of the reaction.

The reaction was completed after the reaction vessel was carefully sealed (with clay) to limit the amount of oxygen as well as the losses of material (by evaporation or combustion) and any potential contamination. The temperature was raised to a value which remains vague in the *hermeneias* [16, 17, 36]. Because it was not necessary to go higher, it is probable that the temperature for the transition from the allotropic form β HgS into α HgS (380 °C) was not greatly exceeded. Chemical and mineralogical analyses should show the appearance in cinnabar of chemical impurities such as mercury sulfate ($HgSO_4$) and mercury, due to oxidation and/or decomposition [1, 40]. In all the Romanian manuscripts, as in the Athonite ones, it was recommended for this step to have a 24 h duration. In the Arab manuscripts the value varies very much from one manuscript to another, and durations anywhere between 12 and 72 h are mentioned.

The last step of the process is the same in all the manuscripts used by the Romanian iconographers [15, 21]. Due to the strong adherence to the walls of the reaction vessel and to its shape, it was impossible to harvest the red product without breaking the vessel. Subsequently, the pigment was scraped off the shards.

The *hermeneias* do not make any recommendations concerning the purification of the pigment obtained. This is different from the manuscripts originating in the Middle East where recommendations difficult to accept and to follow nowadays appear.

In a single Romanian text, reference is made to washing the pigment with water and decanting the supernatant by siphoning with a wool thread [17]. Here we find another proof of a remarkable technical ability which today does not appear as necessary as it did once. Note has to be made that Costin Petrescu[9] and other painters and keepers of the unwritten tradition transmitted from master to apprentice, in their works, claim that by following this recommendation a particular "shine" was afforded to the pigment. When making such a statement, both referred to the subsequent use of the pigment in *fresco*.

The Romanian *hermeneias*, and similar contemporary texts or earlier ones from other parts of the world, only mention two allotropic forms of cinnabar: the red one (α HgS, hexagonal crystals) and the black one (β HgS, cubic crystals). The first is by far the most used and described. This is why when the word chinovar (cinnabar) appears it always means the red pigment.

The photosensitivity of α HgS and the gradual darkening of the color are clearly mentioned in the painter's manuals. When the recommendation against using cinnabar in exterior frescoes (common in Northeastern Romania) appears, it is explicitly stated that this is because "…it blackens" (Chap. 9). Competently, the same text recommends using, in this type of work, a certain combination of pigments (oxiu) made by mixing a reddish clay (bolos) with sinopia[10] or burnt sienna[11] and carbon black (chinoros). It is easy to see that all the pigments found in the composition of oxiu are resistant to both the action of solar radiation and atmospheric agents (Fig. 3.2).

---

[9] Romanian iconographer (1872–1954).

[10] Pigment that was mined in Cappadocia, and exported to Europe through the port of Sinop. Dark reddish-brown due to hematite content (Fe_2O_3, rhombohedral). Widely used in Antiquity and the Middle Ages for painting, and during the Renaissance for underpainting in a fresco.

[11] Sienna is a yellow-brown earth which becomes red-brown after being burnt (hence the name).

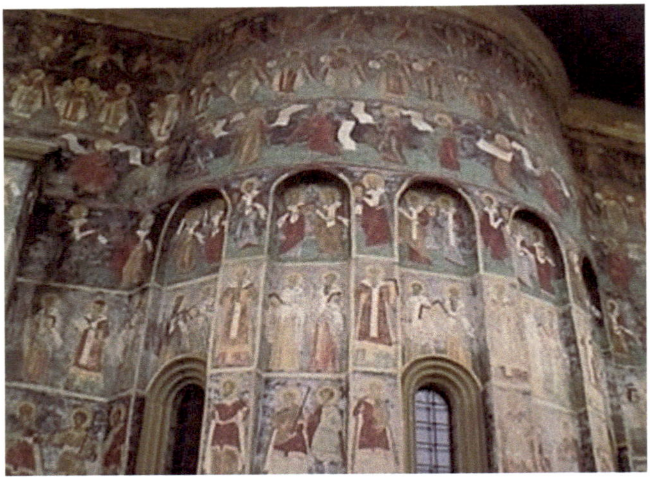

Fig. 3.2 Oxiu, used instead of cinnabar in an exterior fresco (painted in 1601), at Monastery Sucevita (Northeastern Romania). Photo: M.D. Leonida

For the interior frescoes cinnabar was used sometimes mixed with lime white (wall psimith) and ochre (*Tarigrad ochre*). The mixture was resistant to the atmospheric agents and had a hue close to that of oxiu which conferred a chromatic unity between the exterior and the interior murals.

Concerning the incompatibilities between cinnabar and some pigments, examination of its mixtures with lead compounds gives an idea about what was known in this area and suggests the compilation character of these manuscripts.

In some places the interdiction to mix cinnabar with lead white (ceruse) appears clearly. Indirectly, this is accomplished in the texts we discuss herein by recommending to mix cinnabar with lime white. A *hermeneia* recommends directly not to use finely ground cinnabar from commercial sources because it contains minium. Still the situation is not very clear since there are manuscripts which recommend the use in some places (the color foundation for the faces and lips of figures and the shadows on their clothes) of mixtures of cinnabar and lead white [17, 21, 36]. It is true that in these cases the ratio of lead white to cinnabar was very small which made for reduced darkening effects which only appeared after very long times.

It is also true that, as we already mentioned, in other mixtures (e.g. oxiu) cinnabar was replaced by other red pigments, less sensitive and less inconvenient. In one manuscript another color foundation is recommended for use on the faces of the figures, one which was made by mixing cinnabar with another lime-based pigment.

Very seldom was cinnabar used as such, unmixed with other pigments. Some texts recommend the use of cinnabar alone to color the mouth and one of the rainbow bands.

Cinnabar was also applied as a glasiu[12] on top of a minium stain (on red cloth) or, at other times, on top of a very resistant iron oxides-based pigment (chilermineu[13]) to color garments too. In the former case a reduced darkening effect was noticed after a very long time, in the latter, cinnabar was well protected and impeccably preserved.

In red inks cinnabar was used as the only pigment, in aqueous solution. It was first thoroughly mixed with wine, vinegar or brandy. To this solution sugar, gum arabic or myrrh was added [15–22].

3.2 About a Green Pigment, Acetic Fermentation, and Quality Control

As we already mentioned, green vivid colors were discovered much later by those living around the Mediterranean Sea. Greens were malachite—basic copper carbonate $CuCO_3.Cu(OH)_2$ [41], verdigris—$CuO. 2Cu(CH_3CO_2)_2$ [8], synthetic chrysocolla—hydrated copper silicate $CuSiO_3. 2H_2O$ [41], and terre verte—glauconite, a mixed silicate of potassium aluminum and iron, $KMg(Fe, Al)(SiO_3)_6.3H_2O$ [41].

While the Greeks and Romans introduced very few new pigments, verdigris was one of them and white lead was the other one [42]. Theophrastus [8], Vitruvius [43], and most of all Pliny the Elder [44] described in their works the main sources of raw materials for pigments and the technologies to prepare them. Verdigris was among those described therein. Besides written testimonies, archaeological discoveries, especially those at Pompeii and Herculaneum, are important. There sealed jars and bottles containing pigments were found. Being protected from air and light since the year 79 AD when the eruption of Vesuvius took place, the pigments were examined in what should be their original condition.

In the medieval workshops of the iconographers of Southeastern Europe, among the pigments they used, there was a particular green pigment used for coloring clothing, some plant parts and, most of all, for green fields. This is why in the Romanian *hermeneias* the pigment was named "pasture green" [45].

Chemically, the green pigment mostly used by these artisans was hydrated copper acetate $Cu(CH_3COO)_2. H_2O$, having clinorhombic (monoclinic) crystals of a dark bluish-green color. Very close chemically to verdigris[14] (Fig. 3.3), it is not affected by light but it is toxic and not very reliable in paintings because, under the aggressive action of internal (other pigments) and external factors (gases, mois-

---

[12] Shiny top layer.

[13] Earth color, reddish hue, containing clay and iron oxide (syn. *bole*).

[14] Green pigment obtained either through the action of acetic acid on copper plates or as patina when metallic copper, brass or bronze are exposed to air or seawater over a length of time. Hence, chemically, it can usually be basic copper carbonate,basic copper chloride (near the sea), or copper (II) acetate.

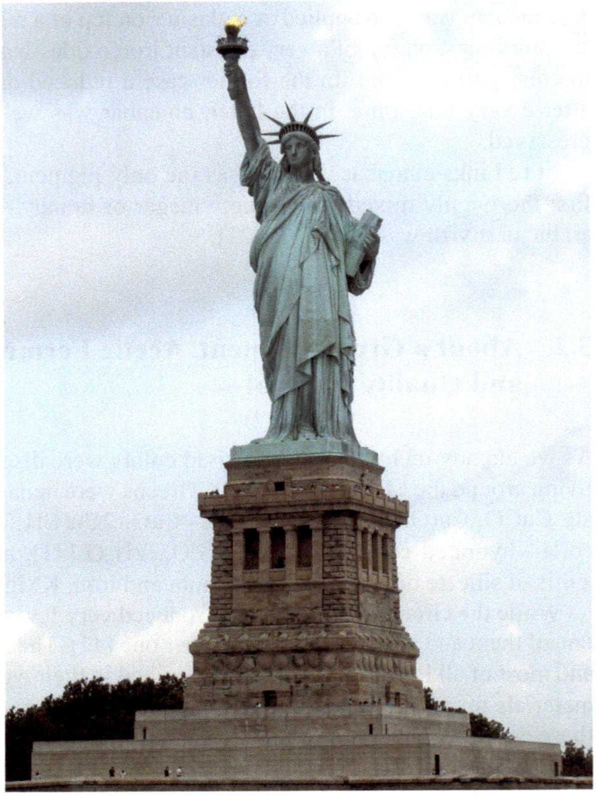

Fig. 3.3 The Statue of Liberty showing verdigris formed due to exposure to the elements and the saline environment on the copper "skin" of the statue

ture) it is unstable and it is converted to other chemical compounds with different hues. Chemical analyses done in such cases may lead to confusing or even wrong conclusions.

In the geographical area we study, this pigment was known under several names. This may provide valuable information regarding economic and cultural relationships at a certain time within that part of Europe. In the Romanian *hermeneias*, for instance, several names are used for this pigment. Some of them were coming from other languages spoken in the area (through technical influences), others originated from names used in countries from where it was imported. Always these names were transformed by the popular etymology. The common sources of these names were the Turkish name ţinghiar, the Italian one verdarame, and the German Grünspan [46]. In all the East-European languages terms were used suggesting that the pigment was looking like copper rust or that in its composition copper was included. Thus, in the Romanian area it was also known as "brass powder" or "brass green" [47].

There are cases when, even in the same technical manuscript, several different terms appear for this pigment. For instance in [16] two different names are used for this green when its preparation is described. One comes from the Turkish name (ţinghiar) and the other is the equivalent of "brass green". A similar situation can be found in the well-known Greek *hermeneia* of Dionysius of Fourna [13]. In all cases it becomes apparent that the author of the manuscript wanted to make clear beyond any doubt which material he was referring to. This is true for the cases when the same pigment was known and used under an old name as well as a new one which had appeared following changes in the socio-economic situation.

Most of the green pigments used by ancient artisans were copper-based although some green earths and green plant juices were used as well. "Brass green" was known, prepared and used in Europe and the Middle East since Antiquity. It was mentioned starting with the old works of Pliny the Elder [44], Vitruvius [43], and Dioscorides [48]. It was also discovered in murals in ancient Egypt and Rome.

This pigment appears frequently in many of the technical manuscripts of the early Middle Ages too. Among those we can mention the treatise of Theophilus Presbyter [34], the standard work on early Northern European methods and materials for painting, that of the Abbasid alchemist Abu Bakr Muhammad ibn Zakariya al-Razi[15] [49], and the famous Lucca manuscript [50]. This is the period when the pigment started to be used extensively in painting and for miniatures and letters at the beginning of chapters in some illuminated manuscripts. While for painting it was used without being mixed with other pigments, in inks and colors used in manuscripts it was sometimes mixed with vegetable saps.

In medieval Western Europe the composition of "pasture green" was different in different periods and also from one place to another because different raw materials, not just copper, were used [51]. This is the reason for its different behaviors over time or when exposed to other pigments with which it was mixed in paints or in inks.

Generally speaking, the behavior over time of this pigment was unsatisfactory. Cennini [10] says about it that it is "very lovely to the eye, but it does not last". Under the aggressive action of internal or external agents, surfaces painted in a lovely green and pages of manuscript decorated in green colors became a dark mahogany brown [51]. This green pigment lasts only under favorable conditions. It is attacked by prolonged exposure to moisture, it is blackened by alkalis (used in cleaning products) or gases in the atmosphere and it reacts (unfavorably) with other pigments.

The Flemish masters' strategy against this was to "lock" the pigment in some preserving vehicle. They either mixed the pigment with a varnish or covered the section of the painting colored in it with a layer of varnish. Similar attempts were noticed in the Romanian area too where icons painted with tempera colors (verdigris included) were covered either with a film of siccative linseed oil or with a varnish. Gradually, after the appearance of more resistant green pigments, "pasture green" was used less and less. In Southeastern Europe, Romanian and Bulgarian painters continued to use it to the end of the nineteenth century.

[15] Persian alchemist, physician, philosopher, musician who described several chemical reactions and about twenty instruments used in chemical work. He was the first to prepare sulfuric acid together with other acids, and he also prepared alcohol by fermenting natural carbohydrates.

The preparative technologies for this copper-based pigment used by the icon painters of the Orthodox East were basically the same [17, 22, 52]. Differently from their West European counterparts, in the East only two raw materials were used: copper and acetic solutions (vinegar).

In a copper pot, concentrated vinegar was put in contact with small pieces of metallic copper. Covered, to prevent mechanical contamination by insects or suspensions from the air, the reaction vessel was left in a place with good exposure to the sun light [16, 17]. These two Romanian *hermeneias*, together with a Greek one [23] state that work was done preferably during days when the air temperature was high. Higher caloric input (heating on a fire) was not used because the centuries old experience of these artisans had recorded correctly the thermal instability of the pigment and had accordingly adapted the preparative technology. The conditions used are also an indirect proof that the painters knew that the higher the reaction temperature, the higher the rate and the yield would be.

The fact that the recipe used by the Southeast European painters can be found in the well-known "Lucca manuscript"[16] too (written many centuries before the *hermeneias*) proves, once again, the influence of common sources over the entire Europe, not just Byzantine influences in the neighboring area, as it is erroneously stated sometimes [47].

The duration of the reaction was not stated clearly in the manuscripts, as it is done today. In the *hermeneias* it was recommended to continue the reaction until "the vinegar will curdle" [22], that is until the viscosity of the newly formed copper acetate solution increased very much, corresponding to a high concentration.

The concentration of the solution was continued in a vessel with large diameter ("wide pot"), by evaporation [16]. However, in none of the *hermeneias* was it stated for how long a time this had to be done. Still, the recommendation concerning the shape of the vessel is proof that the advantage offered by a large surface in evaporation was noticed during the centuries-old technical experience, recorded, and transmitted correctly through the popular tradition.

The copper acetate solution obtained using this procedure was generally kept and used as such in water-based colors for painting on wood panels, on textiles, paper, parchment, and mother-of-pearl. The same solution was used for the preparation of green inks. For oil colors, the concentration by evaporation of the "pasture green" solutions was continued until crystals were obtained. The crystals were subsequently collected and homogenized by grinding until the desired granulation (for a certain type of paint) was achieved. This was always done in mortars made from inorganic materials (agate, porphyry, basalt).

Concerning the size of the copper pieces used as reactant, all the painter's manuals of the Orthodox East require small pieces of copper to be used in the synthesis [17, 22, 23, 47]. Popular tradition correctly recorded that increased surface of reactants favors high rates of reaction and increased yields.

[16] One of the manuscripts included in *Mappae clavicula*, a compilation of recipes for different crafts including painting. The name of this manuscript comes from the town of Lucca where it was compiled (around AD 800) and where it can be found (in the Biblioteca Capitolare Feliniana of the Archdiocese of Lucca).

Differently from Western Europe where sometimes bronze or brass (copper alloys) have been used instead of metallic copper to prepare verdigris [51], in the Romanian workshops only copper was used in the reaction. If the copper contained some lead, the product was contaminated with some white lead produced along with the verdigris. The presence of other metals would certainly change the hue to even a greater extent. In some procedures described by Daniel Thompson [51] the use of copper oxide ground with vinegar was mentioned as well. Such a process would lead to a bluish green pigment with the formula $CuO.\ 2\ Cu(CH_3CO_2)_2$. By using only copper in the synthesis the purity of the pigment was higher, the color of the pigment coming from different batches was the same and its behavior over time was unchanged from one painted work to another. This is why in the paintings of the Romanian painters there are fewer variations from true green to bluish green, and less severe degradations to brown than those mentioned previously.

From the same technical manuscripts and from information handed down through generation after generation of iconographers, it appears that, as a source of acetic acid, only vinegar from wine was used in this area. Sometimes in Western Europe in the synthesis of verdigris marc was used instead of vinegar [10]. In this case the pigment was likely to contain copper tartrates as well as acetate hence its hue was different. When apple vinegar was used, as often was the case in England, other salts, malates for example, may have been present and altered the color [51].

The information about this copper-based green pigment found in the old manuscripts written by painters and meant for their secretive, closed circle stood the test of time. Copper acetate is prepared and crystallized from solution in, basically, the same way in today's laboratory. The pigment is not used in painting anymore because of its toxicity and unfavorable interactions with many other pigments.

3.3 White Pigments Prepared by Iconographers and Used in Mural Painting

With the rise of the civilizations around the Mediterranean new white pigments were discovered. These were chalk (highly pure calcium carbonate, $CaCO_3$), gypsum (calcium sulfate, $CaSO_4.2H_2O$), huntite (calcium magnesium carbonate, $CaCO_3.3MgCO_3$), and white lead (lead white, basic lead carbonate, $2\ PbCO_3.Pb(OH)_2$) [42].

There is no record of use in the area under study, in mural painting, of the white clay used by the Mayan artists (attapulgite or palygorskite, originating from a natural limestone sink-hole existing near the Mayan ruins of Uxmal). The same situation occurs for the white pigments used early in the Far East (China, Japan), kaolin (China clay) or the white pigment made from oyster shells (calcium carbonate-based).

In Western Europe science and art emerged from monasteries (the only place where empirical science and art were systematically practiced) in the twelfth century. Around the fourteenth century the apothecaries became the main pigment merchants. The white pigments sold there and used during the Middle Ages for murals were prepared from limestone, lime, chalk, white marble (a metamorphic form of limestone, calcium carbonate), egg shells and oyster shells [53].

The white pigments used by the painters (religious or lay) of Southeastern Europe for *buon fresco*, chalk *fresco* or *secco* (see Chap. 9) were of four different types, all of them based on calcium compounds[17]. The differences were in the raw materials used, in the preparation, and in the percentage of calcium carbonate present at the time of their use. In some of the pigments calcium carbonate was obtained over time, through the action of carbon dioxide from the air on slaked lime. When water in slaked lime evaporated, calcium hydroxide, its main component, is converted to calcium carbonate following the reaction (3.5):

$$Ca(OH)_2 + CO_2 \longrightarrow CaCO_3 + H_2O \tag{3.5}$$

Because in frescoes color is applied to freshly laid plaster, after having been exposed for a while to the action of atmospheric agents, the chemical composition of all white calcium based pigments becomes identical. That is why it is very difficult, after many years, decades, or even centuries, to recognize which were the raw materials used in their preparation. Eventually all become solid calcium carbonate, chemically the same as the original limestone, chalk or marble.

The most reliable and direct information concerning the raw materials and the preparative methods of the white pigments used in the Southeastern European mural painting is found in the Greek [23, 54, 55] or Romanian [15, 16, 17, 20, 21, 22, 56] *hermeneias*.

Information provided in them mentions four white pigments used during medieval times and later in this area. Two of them contained, at the time of their use, almost exclusively calcium carbonate. Since in true fresco colors are absorbed into wet freshly laid plaster, all calcium-based white pigments, after the painting had been completed and the mortar had dried, will be trapped in the crystalline structure. When, in time, all the water evaporated, the part which was not calcium carbonate yet (but lime—CaO or slaked lime—$Ca(OH)_2$) began to be converted to calcium carbonate by the carbon dioxide in the air (eqs. 3.5 and 3.6).

$$CaO + CO_2 \longrightarrow CaCO_3 \tag{3.6}$$

Eventually all such white pigments become solid calcium carbonate, chemically the same as the original limestone or marble. Carbonation below the surface may take years to complete (Chap. 9). The fact that the painting becomes an integral part of the wall accounts for the good mechanical resistance of frescoes (to air currents, impact with hard particles, etc.).

[17] While in *buon fresco* the vehicle for pigments was clean water, for chalk *fresco* limewater, a solution of lime, was used for that purpose. Because lime is only partially soluble in cold water and even less so in hot water, limewater was used as a very low concentration solution.

Three of the aforementioned white pigments were prepared, through unitary operations, in the painters' workshops while the raw material for the fourth (chalk white) was obtained either from local or foreign commercial sources.

The raw material for two of these pigments was a calcium hydroxide-based paste, lime putty. The Romanian manuscripts call both these pigments lime white. One of them was prepared from calcium oxide (lime, quick lime, a strong alkali) kept for a long time in contact with water (slaking) when the reaction is (3.7) and slaked lime (calcium hydroxide) is produced [57].

$$CaO + H_2O \longrightarrow Ca(OH)_2 + 15.2\ \text{kcal/mol} \tag{3.7}$$

It is this region's version for the preparation of a pigment used during the Middle Ages in Italy under the name bianco sangiovanni [10]. The difference is in the reaction (contact) time, much longer in the Southeast, between six months and twenty years [58]. The long reaction time seems to improve considerably the plasticity and to decrease "popping" effects in the resulting material, the lime putty. When long curing times were used, irrespective of how much longer than six months, the aged slaked lime (which became stiff, paste-like) was called old pit lime putty.

In Italy, during the Renaissance, painters were aging slaked lime in casks stored in pits dug below the frost line; such materials are said to have been handed down through several generations [47]. The pits (where reaction and curing were taking place) used in Southeastern Europe, and described in the *hermeneias*, were 1.5–3 m deep, sometimes lined with wooden planks and always covered (to avoid contamination of the slaked lime). On top of the wooden cover) soil was placed during the storage.

The beneficial effect of long contact times in quick lime hydration was known prior to the Middle Ages. The Elder Pliny recommended in his writings [44] to use lime putty after three years. Vitruvius recommended at least two years, while mentioning that some of his contemporaries obtained very good results with slaked lime after a contact time of two decades [43].

There are no indications that in the Southeastern European area continuous removal of the carbonated lime was performed periodically as was recommended in Russian manuscripts (the periodicity of the process was connected either to numbers with occult significance or to the great religious holidays of the year). Some of the *hermeneias* recommended removing the thin "skin" of calcium carbonate off the lime white surface when the latter was taken out of the storage pit to be used, but not periodically during the first six weeks of storage, as was done on the territory of today's Russia.

In the case of the second lime white calcium hydroxide is the starting reactant too, but it is obtained using shorter contact times. This material was only used for leveling. In the language of the artisans the mortar prepared using this type of slaked lime was called black mortar and was never used in frescos.

Slaking of lime in the case of the first type of lime white was using for reaction (3.7) an amount of water larger than the stoichiometric requirement, about 1 CaO: 2.5 H_2O. The whole amount of water was added from the start. The excess water was meant to make up for the losses due to evaporation over long contact times as well as for the losses in water due to the exothermal effect of the reaction. The purpose of introducing the whole amount of water from the beginning was twofold. It was also meant to improve the quality of the paste obtained following slaking and curing. The calcium hydroxide obtained was to a large extent in colloidal state which means better plasticity and more facile application. The centuries-old experience correctly recorded in the collective memory and transmitted down through generations that an additional amount of water is beneficial and provides also the amount of water the colloidal gel of the future calcium carbonate will include in its own bulk.

There is also another advantage to this first type of lime white. Long contact times make the probability of having unreacted CaO almost zero. If some were left unslaked, when the product is used for fresco mortar, it will be hydrated by the atmospheric moisture after having been included in the wall structure. The resulting calcium hydroxide has a larger volume than the oxide and will give what the eastern painters were calling "shots" and we call "popping": small pieces from the pictorial layer coming loose (see Chap. 9).

A reference to the aforementioned plasticity is clearly made by the recommendations in the *hermeneias* when appearance was one of the quality control criteria used for the process. In some Romanian manuscripts it is recommended for the aged calcium hydroxide to be clay-like [20, 56] (the same recommendation can be found in the Greek *hermeneia* of Dionysius of Fourna [13]) or loam-like [54]. In other Romanian manuscripts, as in similar Russian ones, it is recommended for the slaked lime to be "like pot cheese", like butter, or to have the proper consistency "to be taken by shovel", or "not to fall off the trowel when stirring it". In Cennini's instructions for painters the same type of recommendation appears, to use lime which was "so well slaked that it has the appearance of an ointment".

Another quality control method recommended in the painter's manuals for the aged slaked lime (lime putty) which was the raw material used for both types of lime white was tasting: it was supposed to be neither bitter nor sour.

In the case of the lime putty obtained using short contact times, except for use in brick constructions or leveling, the painter's handbooks advised to use it only if old pit slaked lime is not available.

The third type of raw material recommended by the *hermeneias* and used by the artisans of the Southeastern Europe in past centuries was mortar from very old frescos. There is no test indicated in the painter's manuals for the quality of this material. The long time passing between the completion of the original fresco and the recycling of its mortar was a guarantee that carbonation of calcium oxide (Eq. 3.6) and that of calcium hydroxide (Eq. 3.5) had been completed.

Finally, the fourth raw material used in white pigments in this geographical area was chalk [7]. It was found and mined as high purity calcium carbonate or imported. It has only seldom been used by the painters of the Orthodox Europe and only at later times (last 2–3 centuries) than the first three pigments. Its use was not only for pigments but also as filler in mortar used for the upper layer of a fresco, the "face". The purpose in this case was to obtain a smooth, more luminous surface. It is known that, in order to strengthen the wall, the different coatings of lime plaster are reinforced by diminishing amounts of filler materials. The lower layers have higher filler to lime ratio, while the upper layers are richer in lime. None of the filler materials should have any impurities that could fluoresce and chalk obeys this rule. Use of chalk in this area seems to be an influence from the medieval Western Europe [46].

The use of this last raw material is older than that of the other three. It has been reported in cave paintings, in Minoan murals [59], those in ancient Greece and Rome [60], in frescos at Pompeii [44, 58, 61].

The processing of the aged lime putty was simple enough. After quality control (appearance and tasting) was done, paste or small cakes made from it were air-dried (on planks). During drying the percentage of calcium carbonate was increasing under the effect of atmospheric carbon dioxide. After drying the material was ground and sifted several times.

The processing, as recommended in the *hermeneias*, is different than that used in Western Europe during the fourteenth and fifteenth centuries. Cennini recommended washing the lime putty, after drying, eight times during eight consecutive days, with clean water. The water was removed daily, by decantation. After repeated washing, the calcium hydroxide which had been solubilized was removed as what Cennini called “lime grease”. After washing the process was basically the same as above. Small size cakes were dried in the sun, then repeatedly ground and sifted until the desired granulation was achieved [10]. There is also a difference between the two parts of Europe concerning the grinding. In the West it was done by stone mortars while in the Southeast it was done by stone as well as in wood mortars (see Chap. 8).

Everywhere in Europe grinding was done in clean water for pigments used in *fresco*. Part of the washing procedure mentioned by Cennini was continued during grinding. The white pigment prepared through this procedure has roughly the same composition as bianco sangiovanni. The difference is that while bianco sangiovanni is oxidized lime putty, having no longer a binding effect, the lime white produced from lime putty is pure slaked lime used in making the wall; it acts as a pigment but also as a binder.

For the case when non-aged lime putty was the raw material, after tasting was performed, it was dried in an oven for a long time. After cooling, it was successively ground in the presence of water and dried until the final granulation was obtained and finally dried again.

The old lime putty needed to be processed through more numerous steps than the first two raw materials in order to become white pigment. First the previous colored layer was removed by scraping. The pieces of mortar were then ground and clean water was used to remove the fillers (Chap. 9), using repeated sedimentation and decanting as well as filtration. Some soluble impurities were removed during this process as well. After the mechanical and physical operations, the material was air-dried and then brought to the desired granulation.

The fourth white pigment used in mural painting in the medieval Southeastern Europe was simply prepared from natural chalk, commercially obtained, by successive grinding and sifting.

3.4 Black Pigments

Like painters of the prehistoric times, Antiquity and the Middle Ages, the iconographers of the Southeastern Europe used extensively black pigments. These pigments, predominantly carbon-based, were prepared in the painters' studios from different natural sources. The deep knowledge and artistic skill reflected in the recipes compiled in the *hermeneias* is remarkable to the student of these books. To paint the pupils in portraits for instance, where the Athonite Dionysius of Fourna was recommending only one pigment obtained from fir wood soot, the Romanian manuscripts were recommending several black pigments included in complex color mixtures.

Carbon, in many forms and from different sources, was the most used black pigment not only in this geographical area but universally during the Middle Ages. These materials were obtained by the incomplete combustion of some plant components, of fruit rind, pits, and bark or wood of different wood essences.

Two methods were used to conduct the incomplete combustion:

- deposition of carbon particles (soot) from the smoke produced by the incomplete combustion of an organic material when the flame was directed toward a cold surface (metal or earthenware). These particles, under the name lampblack, were extremely fine in grain and did not need subsequent grinding. They were also used, mixed with gum arabic and water, in inks (Chap. 7).
- Incomplete combustion of organic matter produced a material used to make "charcoal" for drawing prior to applying the paint or, after repeated grinding and sifting procedures, as black pigment.

Another type of black pigment was obtained from natural sources as well. Oak trees, under the action of certain insects, develop swellings called oak-nut galls. The tannic and gallic acids contained in them can be extracted with water. Mixed with solutions of iron salts, a dark purple ink-like compound is produced which gets darker in time. This type of ink was used in Western Europe too. PIXE analysis[18] of the inks used in the Gutenberg Bible found, besides carbon-based India ink, ink made of crushed oak-nut galls mixed with *copperas* [39].

Charcoal, one of the oldest drawing materials was and is still produced by carbonizing splinters of wood. On the territory of today's Romania it was made from local wood essences (willow, hazel tree [16]) as well as from myrtle wood (*Myrtus communis*), just as at Mount Athos and in Western Europe [62]. The wood was dried and cut with a saw and then carved into "pen-like" sticks. These pieces of wood were then placed in a ceramic container. When full, the vessel was covered with a piece of cloth which was wetted with a suspension of clay. After the cloth was dry (thus making a seal), the container was placed in an oven where the wood used as fuel was already in embers ("when half-ignited", "when half-burned" [17]). At the beginning the "pens" would burn with a flame. When the flame subsided, the

[18] Particle induced x-ray emission (PIXE), is a non-destructive elemental analysis technique used for dating and authenticating art objects and archaeological findings.

ceramic container was immediately taken out of the oven, covered in cold ashes or dirt and left to cool. The charcoal was not uncovered sooner to avoid contact with the air and complete combustion.

As in Western Europe [10], a higher quality charcoal was produced by the Romanian artisans from young shoots of grape vines (vine charcoal).

A rapid preparation of charcoal recommended in the Romanian and Greek *hermeneias* was for the cases when the painter "was in a hurry" [20]. In this variant, the small pieces of wood were first wrapped in cloth or paper which was then secured with thin wire. Next, they were placed in hot ashes and left there until the wood stopped burning and no further smoke was produced. At this point they were taken out and covered with cold ash, sand or dirt. The charcoal was left like that until completely cooled. There is no reference in the Southeastern European painter's manuals to the use of the method described by Volpato [62]. According to him, slivers of wood were crammed into an iron tube which was sealed with hot ashes, made red-hot, and then immersed in water to cool off.

When black pigments are obtained by the incomplete combustion of vegetal material followed by the deposition of the resulting small carbon particles on cold surfaces, the material obtained (known under different names in different regions, *lampblack*, carbon black, vegetal black) is pure carbon, a light and fluffy powder, the oldest pigment known and prepared by humans. It was always highly appreciated due to its intense color, resistance to physico-chemical agents, and facile preparation (no need to grind and sift). While Cennini emphasizes the last statement concerning this type of black pigment ("... it does not have to be worked up or ground, for it is a very fine color..." [10]), in a more recent publication Joseph Lambert, discussing *lampblack* produced in China through a process that became standardized by the fifth century, states that "The actual manufacture involved grinding and sieving..." [42]. Since the sources of carbon mentioned for that area were by the eleventh to twelfth century the same as in the Southeastern European area (pine wood) or in Western Europe (vegetable oils), and they were processed also by controlled burning to obtain soot, it is difficult to understand the need for these additional steps.

The *lampblack* appeared in answer to the need for an inexpensive and easy to obtain black pigment. In medieval times this pigment was specifically used for painting on metals. The black pigment, obtained directly in powdered form, was mixed into a varnish till saturated. The mixture was next strained through coarse linen. By twisting the edges of the linen, the captured varnish and black pigment mixture is forced through the linen cloth and into the waiting container. The paint prepared through this procedure was used to paint on metal armors, helmets, weapons [39].

Unlike the Western European sources [51], the Romanian painter's manuals do not recommend for this type of pigment the incomplete combustion of beeswax, linseed oil, hempseed oil, olive oil, incense or even tallow. They recommend only wood as starting material. The wood essences were sometimes the same as those used for heating and cooking.

Lampblack mixed with gum Arabic and a little water was used as an ink (what we call India ink, actually discovered in China) by iconographers in this area too. It has been known and used before the ferro-gallic/ferro-tannic inks. It was highly appreciated because of being permanent. It will never fade. In manuscripts where written parts were lost partially, this situation is due to the decomposition of the binder and not the pigment.

When *lampblack* was obtained from the soot deposited by the smoke from the incomplete combustion of larch tree (*Larix decidua*), it was used by the iconographers to paint the eye pupils ("eye lights") of figures in compositions inspired by Biblical history. It was even justified: in the manuscripts: "because if you use the same black as for fields and garments, it will run"[56]. For fields and clothing *lampblack* made from oak wood, walnut shells, or peach pits ("peach bone black") was used.

Another pigment used for painting pupils was obtained through the same method from resin from conifers ("resin ink"). The resin, from fir tree (*Abies alba*), spruce fir (*Picea excelsa*), juniper (*Juniperus communis*), was burned and the flame was directed to a cold surface to deposit the soot. The pigment was used mainly in frescoes (Chap. 9).

Other vegetal sources used in this area to obtain black pigments by incomplete combustion were: fruit pits, fruit, branches, and bones of domestic animals. The process was conducted in well-sealed containers, placed in small ovens or directly in embers. The charcoal was afterwards subjected to alternate grinding (on hard inorganic materials) and sifting procedures until grains of desired size were obtained.

"Vine ink" was a black pigment obtained from the incomplete combustion of vine shoots. Nowadays this material is called "blue black" due to the traces of gray it produces when mixed with other components. The control of the combustion of vine shoots was even more elaborate (to avoid obtaining a brownish material of unpleasant texture). The shoots were packed tightly in bundles in covered pots and baked in a slow oven. The resulting material was used either as sticks for drawing or for painting. For the latter use more processing followed: dry-grinding (milled dry) followed by prolonged wet-grinding (with water) between two hard stones, until a paste was obtained.

This material was known since ancient Greece (*polignot*) together with those obtained by burning other organic materials like linseed oil, almond shells, and others never used by the Eastern European artisans. In medieval European workshops this material was known as *nigrum optimum*.

The Romanian *hermeneias* recommend to use "vine ink" in frescoes [21]. Due to its mediocre siccativant action, like *lampblack*, when used on wood panels in oils or tempera colors it resulted in many cracks.

"Oak ink" was a black pigment obtained from oak bark by incomplete combustion followed by alternating grinding and sifting steps [46]. When used in frescoes, this pigment resulted in a bluish-black color. Mixing with lime white resulted in interesting grays used to paint Jesus' garments, blue draperies, the sky in Biblical scenes, or as a protective intermediate layer [47]. In Europe, during the Renaissance, Raphael was known for his hatchings done in shades of this gray, which made the color of his skies vibrant. Romanian iconographers, during the same period and later, used differently the chromatic qualities of this black pigment. They used it for painting backgrounds and the hair of old saints with gray hair.

Willow charcoal, much used for drawing from medieval times to the present, is very little mentioned in the Romanian *hermeneias*. The reason may be that it is too light in hue to make a really good black pigment.

Another black pigment, "walnut ink" was obtained from walnut (*Juglans regia*) shells subjected to incomplete combustion, in well-sealed pots introduced into low burning ovens. Due to its high chemical stability this pigment was recommended by the Romanian manuscripts for use in frescoes (Chap. 9) because it could withstand the strong alkalinity of the dilute solutions of calcium hydroxide ("lime water") used as vehicle in this technique.

"Peach ink" or *peach black*, another black pigment, was obtained through the incomplete combustion of peach (*Persica vulgaris*) pits. It was highly prized in all areas of medieval Europe, more so than the blacks obtained from ivory or bones [63]. It is known that the denser the raw material the charcoal is made from, the better the resulting black is. Beside peach pits, almond shells were used in medieval Western Europe but not in the Southeastern area. In Cennini's "The Craftsman's Handbook" [10] the black obtained from burnt almond shells and peach pits is characterized as "…a perfect black, and fine…").

Peach black, known since Greek Antiquity, was utilized in the Romanian area until late in the nineteenth century and was mainly used in mural painting (as such or for violets, both light and dark) due to its fine size and good resistance to basic media.

"Bone ink" was a black pigment obtained by the incomplete pyrolysis of the protein from bones of domestic animals. Note must be made that there is no mention in the *hermeneias* about the use of animal horns or hooves as raw material for blacks.

"Bone ink" has low carbon content (15–20 %), about 60 % calcium phosphate, up to 20 % calcium sulfate and other impurities (some of them are water soluble). It was not recommended to be used in murals because it causes efflorescence (Chap. 9). Although light and fluffy, it is still somehow heavier and more compact than the *lampblack*. It is also less black, more towards brown when compared to the "vine ink". While in Western Europe it was extensively used, even by names like Rembrandt, Rubens, Ingres, in the Southeastern European area it was less used.

Its preparation is described in one *hermeneia* [20] as being incomplete charring of clean bones in well-sealed containers introduced into ovens at high temperatures. In another reference the same pigment was called "ink from oven-burnt in well-covered pot bone". We find remarkable the complexity hidden in this economical statement.

A better black, the true ivory black, which was prepared from ivory chips was not mentioned in the Southeastern European *hermeneias*.

The uses of the black pigments described herein were well defined in the painter's handbooks. For the eye pupils, irrespective of the painting technique, most of the time a pigment obtained by incomplete combustion of organic matter was recommended. This was expressed economically but precisely: "and put the eye lights only in black" [20], or "and put the eye powers and lights with clean black"[17].

As an accomplished artist, the author of a *hermeneia* was not recommending only one chromatic solution. Alternatives were offered affording the same hue and the same resistance to physical and chemical agents. For pupils the alternative was "for the eye lights...use thin *mavra* from resin black". In such a case an Athonite painter's handbook [13] was recommending, without elaborating too much, to use fir tree *lampblack*.

For the hair and beards of old characters, in frescoes, a mixture was used of "old plaster lime" (Sect. 3.2) mixed with "some oak black" [20]. When these hairy parts were shadowed, the recommendation was to use only black pigment.

As a background when anatomic parts not covered by garments were painted, a mixture was used: clay-based ochre, mixed with old plaster lime and an unspecified black pigment [56].

The standard greenish color used for painting garments was obtained from dark ochre mixed with walnut black [20]. A standard color, *proplasma*[19], was prepared from basic lead carbonate, ochre ("green earth") and a black pigment. In the same manuscript, another background color was recommended as well, following the practice in Cretan painting workshops, with a lesser number of components (dark ochre and "a little bit" of black) [17].

So far we have only discussed black pigments prepared by partial combustion of natural organic compounds and used in colors or in some standard background colors. Some of these black pigments were used by the same painters as protective layer against the alkalinity of the *fresco* mortar when pigments sensitive to high pH values were used (Chap. 9). One such pigment was *lazur*, a natural copper carbonate, used to paint large areas representing the sky. At the beginning a "handful" was applied in light ochre, then a second layer with sun-dried lime and oak charcoal was applied, dried and then applied again [20].

Beside the mixtures already mentioned herein, black pigments were featured in several others: "pasture green" (copper acetate} and walnut black in the standard color *mavra*, "oak ink" and white lime for dark grays, or "vine black" and white lime when bluish grays were needed, and "peach black" or *lampblack* mixed with old plaster lime (fully carbonated) for lighter hues. When interdictions to use a pigment or a mixture appear in the *hermeneias* they are justified by expressions like "...otherwise the work won't progress...".

[19] Standard color used as background, applied as monolayer or multilayer, in compositions and for faces of characters in portable or fixed icons painted in tempera on wood panels, and in *fresco* and *secco*. In Western Europe was used under the name *art imprimatur*.

References

1. Cotton AF, Wilkinson G, Murillo CA, Bochman M (1999) Advanced inorganic chemistry. 6th ed. Wiley Interscience, New York, p 1350
2. Mihalcu M, Leonida MD (2009) Iconography in the late Romanian Middle Ages: granulometry issues. Studii Teologice, Seria III. V 4:37–56
3. Leonida MD, Mihalcu M (2005) Old copper-based pigment. Bulletin (NJAS) 50(2):15–19
4. Erdogu B, Ulubey A (2011) Colour symbolism in the prehistoric architecture of central Anatolia and Raman spectroscopic investigation of red ochre in Chalcolithic Çatalhöyük. Oxford J Archaeol 30(1):1–11
5. Gajic-Kvašcev M, Stojanovic MM, Šmit Ž, Kantarelou V, Karydas AG, Šljivar D, Milovanovic D, Andric V (2012) New evidence for the use of cinnabar as a colouring pigment in the Vinca culture. J Archaeol Sci 39(4):1025–1033
6. Consuegra S, Díaz-del-Río P, Hunt Ortiz MA, Hurtado V, Montero Ruiz I (2011) Neolithic and Chalcolithic-VI to III millennia BC-use of cinnabar (HgS) in the Iberian Peninsula: analytical identification and lead isotope data for an early mineral exploitation of the Almadén (Ciudad Real, Spain) mining district. In: Ortiz JE, Puche O, Rabano I, Mazadiego LF (eds) History of research in mineral resources. Instituto Geológico y Minero de España, Madrid. pp 3–13
7. Mihalcu M, Leonida MD (1982) Preparation and utilization of cinnabar in the Romanian "books" of painting and in the activity of the Romanian medieval painters. Rev Muzeelor si Monumentelor Istorice si de Arta XLXI:82–86
8. Mayer R (1991) The artist's handbook, fifth ed. Penguin Books, New York, p 42
9. Mihalcu M, Leonida MD (1981) Romanian medieval techniques for the synthesis of HgS. Bul Fiz Chim 5:242–245
10. Cennini CA (1960) The craftsman's handbook. Translated by Daniel Thompson. Dover Publ, Inc., New York
11. Mazzocchin GA, Baraldi P, Barbante C (2008) Isotopic analysis of lead present in the cinnabar of Roman wall paintings from the Xth Regio (Venetia et Histria) by ICP-MS. Talanta 74(4):690–693
12. Serbanescu F (1978) Rev Muzeelor si Monumentelor Istorice-Monumente Istorice si de Arta 2:78–84
13. Dionysius of Fourna (1974) The painter's manual. Translated by Paul Hetherington, St. Vladimir Seminary Press, Yonkers
14. Mayer R (1991) Dictionary of art terms & techniques, 2nd ed. Harper Perennial, New York, p. 257
15. Grecu V (1936) Carti de pictura bisericeasca bizantina. Ed. Glasul Bucovinei, Cernauti
16. Romanian manuscript 1795, Library of the Romanian Academy, Bucharest
17. Romanian manuscript 5769, Library of the Romanian Academy, Bucharest
18. Romanian manuscript 3719, Library of the Romanian Academy, Bucharest
19. Romanian manuscript 1283, Library of the Romanian Academy, Bucharest
20. Romanian manuscript 2151, Library of the Romanian Academy, Bucharest
21. Ghenadie of Râmnic (1904) Iconografia. Arta de a zugrăvi biserici şi icone bisericeşti. Tipografia cărţilor bisericeşti, Bucharest
22. Romanian manuscript 1555, State Archives, Bucharest
23. Didron P, Durand AN (1845) Manuel d'iconographie chretienne grecque et latine. Imprimerie Royale, Paris
24. Gettens RJ (1972) Studies in Conservation. 17, 45–69
25. De Langlais X (1959) La technique de la peinture à l'huile. Flammarion, Paris
26. Tuckert I (1790) Annales de chimie IV, 25–30
27. Hutington JC (1970) Studies in Conservation. 15, 2: 22
28. Kiplik DI (1954) Materialele colorate ale picturii. Eitura de stat pentru literature si arta, Bucharest
29. Third book of Democritus. Syrian manuscript seventh to ninth century

30. Eighth book of Democritus. Syrian manuscript seventh to ninth century century
31. Last book of Democritus, Syrian manuscript, seventh to ninth century century
32. Syrian treatise on alchemy. Mss. Mm. 6, 29, Library of Cambridge University, Cambridge
33. Sun EZ, Sung YH, Sun SC, Song YX, Chinese Technology in the seventeenth century. Dover Publ., New York
34. Theophilus P (1979) On divers arts. Trans. Hawthorne JG, Smith CS. Dover Publications, Inc., New York
35. Gettens RJ, Feller RL, Chase WT (1972) Vermilion and cinnabar. Stud Conserv 17(2):45–69
36. Sandulescu-Verna C (1979) Erminia picturii bizantine. Ed. Mitropoliei Banatului, Timisoara
37. Smith J (1701) The art of painting in oyl. Licensed, Rob Midgely. London: Printed for Samuel Crouch
38. Harley RD (1982) Artist's pigments c. 1600–1835, 2nd ed. Butterworth Scientific, London
39. Mihalcu M, Leonida MD (2009) Din tainele iconarilor romani de altadata. Ed. Eikon, Cluj-Napoca
40. Waley A (1932) Bull. School of Oriental Studies. London University 6:1
41. Nesse WD (2000) Introduction to mineralogy. University , New York, p 442
42. Lambert JB (1997) Traces of the past. Addison-Wesley, Reading, MA, p 319
43. Vitruvius on Architecture. (1931-1934), Trans. Granger F (ed), G. P. Putnam's Sons, New York, p. 114
44. Pliny the Elder's chapter on chemical subjects (1929) Bailey KC (ed). Longmans, New York, p 123
45. Mihalcu M, Leonida MD (1981) Les erminies-les "livres" de peinture roumains-témoignages écrits uniques pour l'histoire de la science. The 16th History of Science Congress (Ext. Abstr.), Bucharest, Aug. 26–Sept. 2
46. Mihalcu M, Leonida MD (2013) The hidden face of form and color, 2nd ed. Ed. Eikon, Cluj-Napoca
47. Mihalcu M (1984) Romanian medieval values. Ed. Sport-Turism, Bucharest
48. The Greek herbal of Dioscorides (1934) Johnson J (ed) University Press, New York, p 98
49. Ploss EE, Roosen-Runge H, Schipperges H, Buntz H (1972) L'alchimie. Histoire, technologie, pratique. Ed. Pierre Belfond, Paris, p. 206
50. Eastlake CL (2001) Methods and materials of painting of the great schools and masters. Dover, New York
51. Thompson DV (1956) The materials and techniques of medieval painting. Dover, New York, pp 163–166
52. Berger E (1987) Quellen und Technik der Freskos, Oele, und Tempera Malerei des Mitelalters nach Quellenschriften und Versuchen. Reprint 2009 of the 1912 edition, Medieval and Modern Times Series, Saendig Reprint Verlag, Muenchen
53. Moro P, Moro L, Philippot P (1977) La conservation des peintures murales. Compositori, Bologne
54. Ouspensky L, Lossky V (1989) The meaning of icons. St. Vladimir's Seminary Press, Yonkers
55. Schaffer G (1855) Das Handbuch der Malerei vom Berge Athos, Trier
56. Romanian manuscript 1808, Library of the Romanian Academy, Bucharest
57. CRC (1992), Handbook of chemistry and physics, 73rd ed. Daid L (ed), CRC Press, Inc., Boca Raton
58. Verona AG (1943) The painting at Neamt Monastery. Tipografia Cartilor Bisericesti, Bucharest
59. Profi S, Weier L, Filipakis SE (1976) X-rays analysis of Greek Bronze Age pigments from Knossos. Stud Conserv 21(1):34
60. Pfhul E (1969) Malerei und Zeichnung der Griechen (1st ed. 1923). "L'Erma" di Bretschneider, Roma
61. Helbig W (1869) Wandgemelde der vom Vesuv veschuteten Stadte. Leipzig
62. Smith R (1997) The artist's handbook. Alfred Knopf, New York
63. Grandon P, Pastor P (1998) Peintures et vernis. Hermann, Paris

Chapter 4
Gilding

Medieval artists valued gold the most among the metals known at the time. It suggests richness, light, power, homage, permanence. When used in religious artifacts, the presence of gold was meant to impart to the object its noble characteristics (and increase the commercial value) but also to suggest the splendor and perfection of the world "beyond". Besides being associated with splendor, power, wealth, homage, love, this metal has a pleasant color, it can be made to gleam and reflect. Being one of the most inert metals, it occurs in pure form in nature and it does not tarnish or become dull. Gold is also probably the most malleable of metals. It can be beaten into very thin sheets (gold leaf)[1] and drawn out into cobweb filament.

Gilding goes back in time. It was used in ancient Egypt and China, and it is mentioned in the Old Testament and by Homer. In Southeastern Europe, gold mines were known and mining done since more than 2,500 years ago. Craftsmanship in gilding developed during the Middle Ages throughout Europe. It was used in manuscript illumination, in painting on panel (Fig. 4.1), on walls, in solid gold backgrounds (*fondo d'oro).* It reached a peak during the nineteenth century, especially during the Victorian fashion, when gilded objects and ornaments (*ormolu*) were a must in exhibited art objects. In many of the icons (fixed or portable ones) where gilding appears, it was done at a later time than the rest of the icon. In this case the patron wanted to show gratitude towards the saint represented in the icon and who had helped him out of a difficult situation.

In Southeastern Europe gilding was done mostly by adhering thin metal leaf (gold mostly, or silver in rare cases) to a surface or by painting the area with a paint containing metallic (gold or bronze/brass) powder. This was the case in Western Europe too, while in the New World electrochemical procedures were extensively used [1]. Sometimes the gold leaf, after applying it, was left as such. At other times, especially in garments and backgrounds, it was further processed, mechanically—by incising different geometric or floral patterns, or by covering it with a colored layer and making incisions in it (which would leave the patterns in gold color).

Gilding was done on different supports: wood, walls/masonry, colored layers, leather, cloth, metal, stone, parchment. The application was done on grounds or

[1] Purity >95 %, thickness, 0.001 mm.

M. D. Leonida, *The Materials and Craft of Early Iconographers*,
SpringerBriefs in Materials, DOI 10.1007/978-3-319-04828-4_4, © The Author(s) 2014

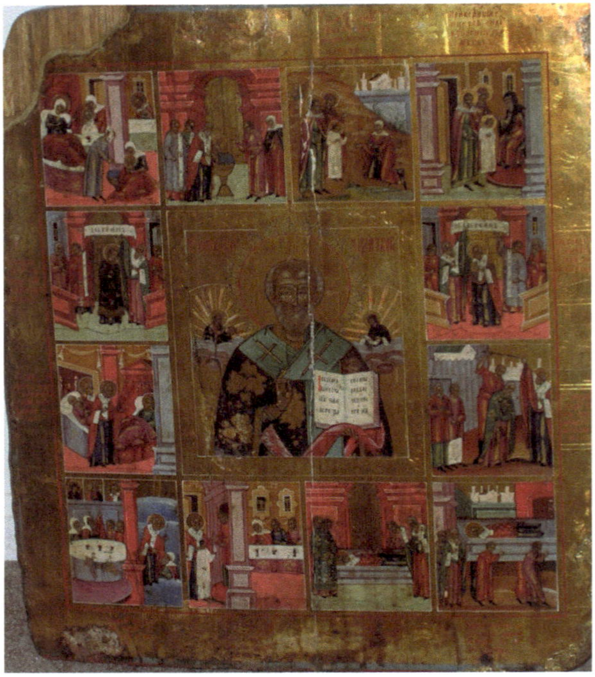

Fig. 4.1 Gold leaf used in Russian icon painted on wood panel (eighteenth century). Photo: M.D. Leonida

directly on the painted layer. In frescoes gilding was done after the mortar was "dry" (the carbonation reaction was completed). In some inexpensive icons gilding was simulated, in haloes, by painting using a yellow pigment (in the appropriate solvent and binder; in frescoes, water or a dilute solution of calcium hydroxide). The painted area was marked out either by a circle drawn in white color or a double circle painted in reddish-brown.

The nomenclature used in connection with gilding speaks about the richness of the technical information the iconographers had, of the "living" book character of the *hermeneias*, about local features, and about connections with the rest of the world [2]. Referring to gilding done on murals, the medieval painters used "to put the gold". "To lay the gold" was the term used for gilding done on wood panels, stone, marble or iron. One term derives from the Russian *polijati* (to gild), while the rest are coming from the Latin heritage of the Romanian language (from *aurum*). A total of eight different words are used in the Romanian painter's handbooks for gilding [3]. Sometimes, trying to avoid repeated use of the same word, different terms are used for gilding in the same manuscript [4].

4.1 Adhesive Mixtures Used in Gilding

For the different adhesive masses used in gilding by early iconographers in the area of study several different names, depending on variations in composition and technical characteristics (adhesive power, color, plasticity, rigidity) after they had been put in place and had dried, were mentioned in the *hermeneias*. Almost all adhesives used in gilding, with the exception of *ilinocopia,* were homogeneous masses, showing plasticity and requiring different periods of time for drying at room temperature. They can be divided, as in Western Europe [5], into water-based ones (*ambole, poliment, ilinocopia)* and oil-based ones (*murdent, gulifarba, mixtion).* In Western Europe all were falling under the term mordant (either water- or oil-based) and their use seems to have started prior to the twelfth century. The names used in Southeastern Europe are an indication of cultural and commercial connections across Europe during the period under study: mordant and *murdent,* or *gulifarba* which seems to come from the German *Gold* (gold) and *Farbe* (color).

Depending on the adherence to different types of supports: wood panel (Fig. 4.1), mortar, masonry, surfaces painted in different media, leather, metal, stone, mother-of-pearl, some of the adhesive masses discussed herein were used on several such supports (ex: *murdent*, having linseed oil in its composition, was known to have good adherence on different supports), while others were used on only one type of support (ex: snail slime which was used only in manuscripts). Recipe components in adhesive masses had different roles and sometimes one of them could fill different functions in one formulation. (Table 4.1)

Some pigments introduced in the adhesive mixtures were meant to modify the hue of the gold leaf beyond what subsequent burnishing would achieve. Through a very thin, almost transparent layer the pigments would temper the color of gold. To "warm up" the color, lead oxide (Pb_3O_4) or cinnabar were used and this is a characteristic of the region. In Western Europe dark red was very seldom used to temper the color of gold leaf. To "cool down", copper acetate (*verdigris,* locally known as *vardalam* [6]) and indigo were used, while burnt sienna (*ghiulbahar*) was used for darkening.

Other pigments used in adhesive preparations for gilding were:

- bole (*Armenian bole*). This was a generic term used for a velvety-smooth reddish earth composed of clay and red iron oxide (Fe_2O_3). It was used in gilding to temper the color and size [7];
- English red (*englirod*)—another pigment composed mainly of anhydrous ferric oxide and used mostly in mural painting;
- massicot[2] (*sulighen*)—yellow allotropic modification of PbO, similar to litharge, also known as Flemish ochre;
- "Icon white" (*ceruse),* basic lead carbonate $2PbCO_3.Pb(OH)_2$.

[2] Lead(II) oxide in an orthorhombic lattice structure, PbO.

Table 4.1 Recipe components for adhesive preparations used by iconographers

Component	Role							
	Pigment	Filler	Binder	Siccativant	Vehicle	Plasticizer	Ammal-gamation	Modify color of gold leaf
Bole	+	+						
Ochre	+	+						
Minium	+							+
Cinnabar	+							+
Copper acetate	+							+
English red	+	+						
Burnt sienna	+							+
Ghiul bahar[a]	+							+
Ceruse	+	+		+				
Chalk			+					
Plaster of Paris			+					
Paper Ash			+					
Bees wax					+			
Tallow					+			
Spermaceti					+			
Soap					+	+		
Linseed oil			+		+			
Egg white			+		+			
Ox bile			+					
Glue			+					
Rock candy			+					
Mercury							+	
Brandy					+			
Water					+			

[a] Pigment obtained by calcination of burnt sienna

Besides tempering the color of gold and enriching the chromatic accords in an icon, some of these pigments (bole, massicot, basic lead carbonate) were sometimes functioning as fillers too. In this case they appeared in a larger amount in the recipe. When linseed oil was used as a binder, lead carbonate and massicot were functioning as siccatives too. A more usual siccative used in adhesive preparations in this geographical area was zinc sulfate ($ZnSO_4$, *white vitriol*).

Other agents used as fillers in gilding were: chalk, paper ash, plaster. Plaster only appears in water-based adhesive preparations, more so in those of the *ambole* type than in those of the *poliment* type. In the Athonite *hermeneias* [8, 9, 10] plaster is not mentioned as part of adhesive preparations used in gilding and the same is the case in different technical manuscripts originating in Western Europe [11, 12, 13, 14].

The raw materials with adhesive properties used in these preparations were: egg white (from eggs not older than 5 days), glue (from either bones or hide) as a dilute aqueous solution, rock candy, ox bile, and linseed oil (as both binder and vehicle)[3].

Other substances used as vehicle were beeswax ("white wax", "holly wax"—probably wax for candles which were meant to burn without smoke, a material for which combustion was complete)[4], beef tallow ("candle wax"), soap[5], and imported spermaceti[6].

Dilute aqueous alcoholic solutions, made with different brandies and water, were used sometimes to decrease the viscosity of the adhesive preparations, sometimes as vehicles in these preparations. Concentrated alcoholic solutions were sometimes used to reactivate adhesive layers previously applied, shortly before applying the gold leaf.

Finally, mercury ("quicksilver") can be found in some recipes of *ambole* or *poliment* to create intermediate layers of higher strength (by amalgamation) between the support and the gold leaf.

The recipe of an adhesive preparation was targeting several properties: as good adherence as possible of the gold leaf to the support, ease of use, time to harden, duration of the application process, beneficial influence on the final color of the gold leaf.

The adhesive agents were prepared in each iconographer's workshop in step-wise procedures. The solid components of the recipe were brought—either together or separately—to a fine ("powder-like") granulation (as appropriate for each type of adhesive). Crushing was followed by breaking up the big pieces, grinding in stone mortars alternating with repeated sifting. After that, separately or together, the granules were stored in paper bags, cloth bags or ceramic containers and amounts as needed were taken out. The fast hardening binders were added shortly before use as necessary for one-time use. Apprentices under supervision performed these mechanical manipulations, while the mixing of the ingredients was always done by the master personally.

---

[3] Linseed oil was obtained by pressing the seeds of the flax plant which is grown in countries with relatively cold temperature. It is superior to all other "drying" oils because of its resistance to embrittlement together with color retention. It was used in adhesive preparations either "unboiled", meaning non-siccativated or "boiled" (siccativated). Siccativation was performed by exposure to sun light, by heating, or by mixing with siccatives like lead carbonate or lead oxide (Pb_3O_4). In the studios of Southeastern European early iconographers it was used as a binder in oil colors, in adhesive masses of the types *murdent* and *gulifarba*, and to impregnate and conserve wood. When it was used in the grinding (in mortars) of pigments with siccative properties, the linseed oil was used "unboiled".

[4] Beeswax was extracted from honeycombs by melting. It was purified by alternating melting processes with cooling by slowly pouring it into cold water. It was discolored ("whitened") by exposure to sunlight.

[5] Soap for this application was prepared from higher quality raw materials (vegetable oils extracted at low temperatures, and the lye was from certain types of wood). It was either a liquid or a paste and its role in the recipe was as an emulsifier, vehicle, and plasticizer.

[6] Spermaceti is a white, translucent wax originating in the head cavity of the sperm whale. It is more brittle than other waxes. It was used in the past to manufacture candles of a higher quality.

The preparative technologies described have regional characteristics. Western Europe and England in particular valued aged (rotten) organic products like glair (a preparation based on processed egg white) because of its increased stickiness. Romanian artisans recommended preparing the adhesive masses in amounts necessary for immediate use, within hours, in order to prevent biodegradation (by air, fungi, bacteria) and/or unwanted polymerizations.

Concerning the actual use of the adhesive preparation the step is mentioned separately from the actual gilding and sometimes it is called with the name of the adhesive preparation itself (ex: to *ambole*). The real meaning of this verb becomes obvious when in a Romanian *hermeneia*, after the preparation of the adhesive is described, the recommendation to subsequently *ambole* and then gild appears [15]. Another *hermeneia* [16] mentions the verb *to poliment* and explains "that is the laying of the gold for the gilding to come out good". Another technical manuscript [17] offering recommendations for gilding on plaster was breaking down the process as "…first lay the plaster, then *poliment*, gild, let it stay for an hour, and then burnish…".

4.2 Ambole and Poliment

These two types of adhesives were prepared practically with the same components and had very similar properties. This situation is mentioned by the authors of the *hermeneias* as well. The title of one such recipe of an adhesive mass in a painter's manual was "*poliment* which is called red *ambole*". A later manuscript, which was a compilation of three others (Sect. XVII–XVIII) together with the lifelong experience as an iconographer of the author, mentioned "red *ambole* which is red *poliment*". The same situation is mentioned by Max Doerner for the painters of Western Europe during that time [14].

Different technical manuscripts treat differently this topic. While in the Romanian *hermeneias* a total of eleven different recipes of *ambole* are mentioned, in the Athonite *hermeneia* of Dionysius of Fourna [8] only three are mentioned. The fact that there are some Romanian manuscripts in which only one recipe is mentioned shows that, even within the Romanian area, as everywhere else, the complexity of the experience and information were different from one workshop to another.

The main components in the two types of adhesive preparations were: pigments—bole, ochre, lead oxide (Pb_3O_4), cinnabar (HgS), burnt sienna, indigo; raw materials with adhesive properties—egg white, oil, ox bile, rock candy; fillers—pigments (most of the time); vehicles—beef tallow, beeswax, spermaceti, mercury, ethylic solutions.

Sometimes the preparation of *ambole* and *poliment* was done at room temperature, sometimes, when beeswax was present, this ingredient was first melted (on low fire) and only then the other components of the recipe were added. From there on two subsequent situations are mentioned. Either the preparation was used immediately, or in the cases when easily biodegradable compounds were absent, small cakes were made that were dried and stored protected from impurities. When they were used, they were softened in water and 2–5 layers were applied on the support. After the last layer dried, immediately before gilding, it was activated with brandy in which a little egg white had been added.

Let us have a look at each category of materials included in adhesive preparations of *ambole* and *poliment* types.

The pigments. Ochre (a clay-based pigment with colors from yellow to deep orange/brown) appears in all *ambole* recipes and only in less than half of those of the *poliment* type. Most of the time its role is as a filler, like that of chalk or plaster. It was recommended for ochre to be of the best quality, without sand and to have the same qualities as *Tarigrad ochre* [4, 18]. It was very common in the technical manuscripts of the Romanian iconographers to characterize and recommend a raw material with reference to another one, better known, from a well-trusted and widely used source. To temper the color of the gold in the same direction as ochre, saffron was used in Western Europe while in the Southeastern part it is not mentioned.

Bole, like ochre was part of more than 90 % of *ambole* formulations and of only less than one third of the formulations named *poliment*. It had a double role, as a filler and to "warm up" the color of the gold leaf. Bole (known in Western Europe as *Armenian bole* or *poliment*) is required in the Romanian *hermeneias* to be "not exactly red", to have "white veins" (caolin) of a lesser density. The hardness was checked using the finger nail [4, 19]. The same recommendation can be found in the Athonite *hermeneia* of Dyonisius of Fourna. Other names by which bole was known in the area were *chilermineu* [15, 19, 20, 21], *vol* [18], and *englirod* [18, 20].

Cinnabar appears in 20 % of *ambole* recipes and in none of the *poliment* type. This pigment was more expensive than minium (Pb_3O_4 which was easier to prepare and denser). This last feature encouraged dishonest merchants to mix minium with cinnabar and to sell the mixture at the cinnabar price (see Sect. 3.1) and this possibility is mentioned too in the Romanian handbooks. Minium appears in over 50 % of the *ambole* recipes and in less than 20 % of the *poliment* ones. Both minium and bole had also the role to "warm up" the color of the gold leaf.

Burnt sienna appears in less than 10 % of the *ambole* preparations and in less than 20 % of the *poliment* ones. Indigo is mentioned in only 10 % of the *poliment* recipes. Both were used to temper the color of the leaf used in icons, although in opposing directions.

The vehicle. Beeswax, obtained by repeated melting, alternating with cooling by slowly pouring in cold water, was either introduced in the preparation (about 52 % of the recipes mentioned it) as such or after prolonged exposure to sun light to whiten it.

Beef tallow appears in about 30 % of recommendations under the name "candle fat". This probably implies an advanced purity resulting from melting followed by separation of the cellular material [4, 15, 18, 20]. The amount required was usually indicated, not always very precisely: in one case it is supposed to be "as much as a grain of lead", in another it is mentioned only as "a little".

The soap in adhesive mixture recipes was used as a paste. It was made from vegetable oils and lye wash obtained by scalding ash from corn cobs, beech or oak wood, then decanting and using the supernatant. As a component of the recipe, soap was used as a plasticizer as well. The names used for soap in these technical manuscripts vary from "Turkish soap" or "brandy soap" [15] to "good soap" [18, 20] indicating that the saponification had to be correctly conducted, using the reactants in stoichiometric amounts and resulting in a soap with reduced alkalinity. In about half of the recipes the amount of soap is not indicated. This is probably because at that time, due to the technologies available, and large differences in raw materials, wide variations in the soap's alkalinity, composition, and consistency were quite common.

Two recipes for adhesive preparations [4, 15] ask for simultaneous use of three different vehicles (soap, beeswax and spermaceti). Neither in the Athonite *hermeneia* of Dionysius of Fourna nor in Cennini's Handbook are there adhesive preparations used in gilding which contain three vehicles simultaneously as in the Southeastern European area. Beeswax is associated with tallow in several of the recipes (about 26 %) and with soap in about 3 % of them. About 44 % of the preparations contain only one vehicle. Note has to be made that in no recipe spermaceti appears alone and also that in about 13 % of the preparations there is no bulk component functioning as a vehicle only.

The adhesive properties of the mixtures used in gilding are due to the presence of egg white or dilute aqueous solutions of glue, and only one Romanian *hermeneia* mentions the use of ox bile too [15], as does a Byzantine one [8].

Glue obtained from bovine hide using recipes given in the same manuscripts (Chap. 5), appears only in a few of the recipes for adhesive preparations for gilding and when it does it is said that it has to be in a dilute solution and the amount to be added is left to the discretion of the person in charge of mixing the ingredients.

The majority (70 %) of the recipes for *ambole* and *poliment* recommend the use of egg white (beaten until frothy, mixed with a little water and allowed to stand until the froth subsides) as was the practice at Mount Athos [8] and in Western Europe as well [12, 21]. The difference is that the Romanian *hermeneias* recommend to use fresh egg white; one of them [19] even specifies that the eggs have to be not older than 5 days while Western European painters let *glair* (egg white and water) become rotten because it becomes stickier. A book on these topics published in England [5] even says that "the older and more rotten they be, the better they be". A strange detail is the indication in the same book to use a little ear wax when glair is used, to prevent the formation of air bubbles in the paste. We did not find such a component in the Southeastern European recipes. The amount of egg white required in some recipes from the Orthodox painter's manuals is indicated exactly [18, 20], in others the recommendation is less precise [12, 18] while in still other recipes no amount is specified at all [15, 18, 22, 20, 23].

Concerning the fillers used in adhesive preparations of the *ambole* and *poliment* type the Romanian iconographers were using, besides some pigments (as previously mentioned), plaster, paper ash in about 45 % of the *ambole* formulations and chalk, plaster, and paper ash in 24 % of the *poliment* type of recipes. The double role as pigments and filler is confirmed by the ratios in the recipes: when one filler is mentioned the amount of pigment is much less than for the recipes with no bulk-making component present.

In Western Europe fewer fillers were used. In Italy slaked plaster of Paris (*gesso sottile*) was used (on wood panels), while in France, England and Northern Europe chalk was the filler of choice. Basic lead carbonate was used as a filler too [5].

The presence of mercury in some *ambole* and *poliment* recipes was meant to increase the adherence of gold leaf by creating an intermediate thin layer of amalgam. This was useful especially when the support gilded was a metal. In old times in Western Europe the object on which mercury had been applied was subsequently heated so that the mercury evaporated and a (two-metal) amalgam was left. In later years this "fire gilding" was replaced by electroplating due to the hazard to people conducting the overlaying operation. Recipes containing mercury are presented in [18, 20] and were used both in the Romanian area and at Athos. In Cennini's Handbook no recipe containing mercury is mentioned. When mercury is required *hermeneias* recommend first to disperse it well into tiny spherules by putting it in one palm and breaking it down with a finger (moistened with saliva) from the other hand.

To dilute the adhesive preparation of the *ambole* and *poliment* type water and alcoholic solutions (brandy) were recommended. No ratios are indicated and previous experience, viscosity, and particular characteristics were the decisive factors in setting working ratios. When beeswax was part of the composition, the brandy was added at the end after the wax had been melted and all the components mixed. Brandy was used also when several thin layers of adhesive had been previously applied alternating with drying steps. It reactivated the adhesive already on the support from previous applications for the subsequent application of the gold leaf.

As a diluting agent and as a vehicle as well, water was also used. Preparations containing egg white, which were stored as small dried cakes were softened in water before use. After the application, the adhesive was allowed to dry and then it was reactivated by wetting it with brandy.

4.3 Murdent, Mixtion and Gulifarba

These types of adhesive preparations were mainly used in gilding done on murals. In one of the texts it is explicitly stated so [19], in another one [4] it is required to use only *murdent* in murals because with *ilinocopie* "...it will later get moldy and rot...". These preparations were used on other supports as well: cloth, leather, marble, wood, oil paint.

Unlike the *ambole* type of preparations, in the case of *murdent* the number of recipes available in Romanian manuscripts [18, 19, 20, 24], in a compilation of a traditional icon maker who lived in the first part of the nineteenth century at Mount Athos, the very heart of Orthodox Europe [8], and in the Russian painters' work [25], is quite small. The small variations from one manuscript to another are not significant from the point of view of the functionality of the adhesive preparation.

In the *murdent* adhesives, linseed oil (boiled and with good siccative properties [26]) enters as both binder and vehicle. In all *murdent* recipes containing linseed oil, the oil makes about 50 % of the preparation, the same as in the recommendations coming from the workshops at Mount Athos.

Raw materials entering the recipes for *murdent, gulifarba* and *mixtion* can be divided into the same categories as those for the water-based adhesives.

The pigments used in *murdent* adhesives were: massicot, basic lead carbonate, minium, copper acetate (not found in the water-based preparations and meant to "cool" down the color of the gold leaf while the others can temper the color in the opposite direction).

The heavy metal oxides included in the recipes had a role as siccatives as well, decreasing the time for the linseed oil to polymerize (to "dry"). In a recipe found in two of the Romanian *hermeneias* [18, 19] a special siccative is added (zinc sulfate). As in the other formulations, when pigments have the role of fillers too, their relative amounts are higher. Only two recipes recommend the use of agents serving only as fillers [4, 25]. Those agents are fine burnt clay powders and shell powder, the same as those used at Mount Athos.

Neft[7], was used as a specific thinner for linseed oil and as a vehicle for the application [4, 18, 19]. Neither the Athonite manuscript, nor the Romanian ones considered it always necessary to actually mention its use.

This type of adhesive preparation was done stepwise and only shortly before using it. From previously mechanically-conditioned components, the amount for one use was weighed, as was the linseed oil. The mixture was then heated slowly (3–4 hours) in a ceramic container until "it becomes darker" or, in another recipe, until "it thickens". When the recipe contained a siccative, this was introduced in the mixture after the heating had stopped.

The application of the linseed oil-based adhesive preparations is extensively described [4] and was done on several supports: mural, varnish layer, marble, leather. It was recommended to expose the object to the air for about three days then the parts which were supposed to be gilded were covered with *murdent* which was left to dry until, when touched with the finger, it showed stickiness without coming off onto the finger. Then a dilute solution of glue was applied, upon which the gold leaf was laid and left to dry for the next 3 days. Burnishing was recommended after drying. When the support is stone the sequence was reversed, glue was applied first, dried, then the adhesive was laid out and partially dried and then the gold leaf was applied and burnished.

In the case of *mixtion* a very detailed recipe was found in the *hermeneias*: 30 parts ochre, 4 parts copper acetate, 8 parts minium, 10 parts massicot, 30 parts basic lead carbonate ("icon white"), 200 parts boiled linseed oil, and 4 parts zinc sulfate. The preparation followed the same steps with the difference in the heating (done quickly) and under continuous stirring until the color darkens. After the heating stops, zinc sulfate is added under vigorous stirring. In the case when gilding was done on a painted layer of an icon, in order to improve the adherence to the support,

[7] Term used either for turpentine or for a mixture of benzene, toluene and xylene.

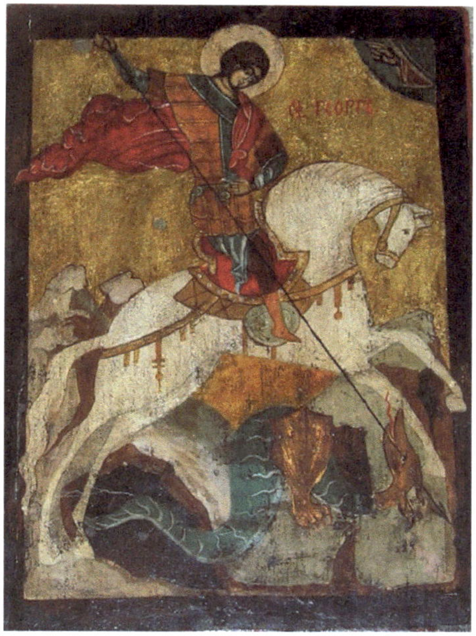

Fig. 4.2 Icon (on wood, eighteenth century), St. George Slaying the Dragon. It shows extensive gilding: in the background and (burnished) in the dragon (head and foreleg). Photo: M.D. Leonida

this was repeatedly touched with a bag made from tightly woven cloth in which either chalk powder or plaster was introduced. After this step *mixtion* was applied.

When the gilded parts are burnished, shadows and lights appear in the reflection as well as lights; as Cennini says, gold "appears almost dark from its own lustre" and instead of looking light and yellow, it looks dark and metallic. The strokes of the burnishing can sometimes be seen in the gilding and were carefully planned and applied by the artisan. In the icon in Fig. 4.2 gilding was done in two areas. While in the background the gold leaf was not burnished and it appears equally shiny over the entire area, in the foreleg and the head of the dragon, due to burnishing, a darker shade is imparted to the gold color and shadows and lights appear. Because burnishing is a smoothing process it can be done only while the adhesive layer is not fully dried.

The smoothing is actually done on that layer, and the gold leaf, pressed on this surface, takes a spectacular gloss. Ralph Mayer adamantly says [27] that burnishing can be done only on water mordant gilding, Thompson seems to agree without so much emphasis [5], while Doerner [14] mentions that in the open air only oil gilding will last while the water-based one can be burnished and used only inside. In South-eastern Europe both types of adhesive preparations were considered fit for burnishing before the drying was completed. Before burnishing the surface was smoothed using a rabbit leg. The tools used for burnishing were a hard mineral, mostly agate, or "the bone", a tool made from an animal tooth. The carnivorous ones were considered superior. In England, for certain less important situations, even a horse tooth was recommended as being bigger and flat.

Oil-based adhesives were recommended by the Romanian medieval artisans for gilding done on cloth as well. First, the parts which were to be gilded were soaked in linseed oil and then allowed to dry for 24 hours. Then the adhesive preparation was applied on cloth, partially dried, and then gilded [25]. In other parts of the world other procedures were used: in Western Europe thin strips of gold were wound around silk yarn (the result had little flexibility) which was then woven. Later the membrane thread process, which appeared in Asia in the eleventh century, was introduced in Europe by the fourteenth century. A binder like egg white was mixed with powdered gold and applied to leather, animal gut, paper, parchment. Narrow strips of these gilded materials were wrapped around some type of fiber to form a thread. The procedure was extremely labor-intensive [1].

4.4 Ilinocopia

Ilinocopia is garlic (*Allium sativum*) juice and it was used by the medieval iconographers as an adhesive in gilding or for golden letters in illuminated manuscripts (when it was mixed with gold powder). Mention of vegetable juices (garlic, sap from the fig tree) used as adhesives can be found in early writings like those of Vitruvius [28], Cennini [12], Vasari [29], as well as in Venetian and Byzantine manuscripts. Most of the time garlic juice is mentioned either alone or mixed with some color and white lead. By the fifteenth century a mix of garlic juice and ammonia became popular for use in gilding. In Southeastern Europe, during medieval times, garlic juice was used as such, or sometimes it was mixed with a little gum arabic and rock candy. In Western Europe sugar or honey was added to *glair* to make it less brittle and keep it from drying too fast. The supports it was used on were portable or fixed icons (on wood), paper and murals. This preparation was sometimes used also (as it was used in Western Europe too) to increase the adherence of oil colors when they were applied on supports of very low porosity.

All authors agree that this adhesive can be used only in places with low humidity and good ventilation. Otherwise the high content in protein (6.5 %) and other organic compounds soluble in water (approx. 0.8 % carbohydrates) make it prone to getting moldy and rotting.

Those who compiled the Romanian and the Greek *hermeneias* insist on the requirement to use big cloves harvested during summer time when the concentration of components with adhesive properties was at its highest. While the Romanian "books" recommend this to be during June-July [4], the Athonite "books" recommend June [8, 9] (see Chap. 2).

The cloves were peeled and then crushed in brass mortars. The product was then squeezed in loosely woven cotton bags and the juice collected in a clean container. Sometimes it was subsequently used as such [21], as it was at Mount Athos and in the Romanian area. Sometimes it was mixed with sugar and gum arabic and kept in the sun [8]. After filtration it was kept in well-sealed containers, and both Western and Southeastern European painters insist on that.

An old iconographer was writing at the beginning of the eighteenth century that the older that juice is, the better it is. In reference [4] two recommendations are found: a) if the juice is too viscous, it can be diluted with water and a little ammonia; b) in the case of garlic juice mixed with gold powder used in writing, it has to be dried completely (by exposure to the sun) before it can be varnished.

The use of garlic juice-based adhesives was very simple: application with a brush and drying to a degree which was supposed to be learned by experience and, for the apprentice, by carefully observing his master. If the drying was too advanced, breathing on the adhesive was recommended to make it sticky enough to proceed further. The gold leaf was then laid over, then cleaned and smoothed with a clean and dry rabbit leg. When garlic juice-based adhesives were used in gilding in Greek and Romanian manuscripts (in illustrations, titles or letters at the beginning of chapters), the areas to be gilded were first covered in fish glue which was allowed to dry before applying *ilinocopia*.

4.5 Gilding on Paper

Snail slime (slaver) was used by the illustrators of the medieval Romanian manuscripts as binder-vehicle for silver or gold powder used in illustrations or letters. The secretion was stimulated (in order to obtain a higher amount in a short time) by introducing a burning candle in the orifice of the shell ("the nostril") [4, 15, 21]. The same procedure was recommended by Dionysius of Fourna with the difference that the stimulation was done by dripping molten wax into the aperture. The slime was then mixed with aluminum-potassium sulfate, $AlK(SO_4)_2$ and gum arabic, in a mortar until it was fully ground and homogenized. To this mixture gold powder was added and again homogenized. The result, "gold ink", was applied where needed, allowed to dry and then burnished. This type of adhesive preparation cannot be found in other geographical areas.

When gold leaf was applied on paper, before applying one or another of the adhesive preparations, several layers of dilute solutions of glue were applied, alternating with drying steps.

Another "gold ink" was made by mixing gold powder with an aqueous solution of glue or gum arabic. The gold powder was obtained by grinding thin gold leaf either in mortars made from a hard mineral or on a thick glass (with no irregularities, air bubbles, or variable thickness). In a clean mortar approx. 10 g of melted beeswax (strained through a "clean cloth") and two spheres made from gold leaf, were ground together until a "butter"-like paste was obtained. The paste was transferred into a glass with water and allowed to stay there for about 8 hours. After that, the water was decanted carefully (not to lose any gold). The grinding was repeated, this time together with the adhesives and the viscosity was adjusted as desired with brandy or water. If this ink was to be used on oil-painted areas, these were previously covered with plaster of Paris. It was not recommended to apply the ink directly on the painted layer. There was no need of plaster if the colored layer was tempera [18].

Considering all aspects, gold powder was more difficult to obtain and considerably more expensive to use than gold leaf.

For writing on murals, a mixture in equal parts of glue, ash lye and beeswax was used as a binder-vehicle for the gold powder. The three components were heated first and only then water and gold powder were added and the mixture was stirred vigorously until homogeneous. After application and drying, burnishing was done [15]. Unlike other water-mordant based procedures, no varnishing was needed when this adhesive preparation was used.

4.6 Gilding on Mother-of-Pearl

This procedure was mainly used for gilding on small expensive icons made either partially or entirely of mother-of-pearl. The procedure was rather simple. First an egg white-based aqueous ground was applied. After air-drying, the religious scene or portrait was impressed on it by scratching. A second layer of ground (containing a small concentration of white pigment, basic lead carbonate) was applied. After drying, the religious scene/portrait was impressed again, this time using colors (tempera or oil) with a high amount of binder. After drying, the gold leaf was fixed with mordant and varnished with a hard resin (copal, sandarac).

References

1. Lambert JB (1997) Traces of the past. Addison-Wesley, Reading p 204
2. Mihalcu M, Leonida MD (1981) Les erminies—les livres de peinture roumains—temoignages ecrits uniques pour l'histoire de la science. The 16th History of Science Congress, Bucharest, Aug. 27–Sept. 3
3. Mihalcu M, Leonida MD (2009) Din tainele iconarilor romani de altadata. Ed. Eikon, Cluj-Napoca
4. Romanian manuscript 1795, Library of the Romanian Academy, Bucharest
5. Thompson DV (1956) The materials and techniques of medieval painting. Dover Publ., New York
6. Leonida MD, Mihalcu M (2005) Old copper-based pigment. The Bulletin (NJAS) 50(2):15–19
7. Mayer R (1991) Dictionary of art terms & techniques, 2nd edn. Harper Perennial, New York, p 45
8. Dionysius of Furna (1974) The painter's manual (Translated by Paul Hetherington). St. Vladimir Seminary Press, Yonkers
9. Didron P, Durand AN (1845) Manuel d'iconographie chretienne grecque et latine. Imprimerie Royale, Paris
10. Schaffer G (1855) Das Handbuch der Malerei vom Berge Athos. Druch und Verlag der Fr. Lintzschen Buchhandlung, Trier
11. Eastlake CL (1967) Materials for a history of oil painting. Dover Publ., New York
12. Cennini CA (1960) The craftsman's handbook (Translated by Daniel Thompson). Dover Publ. Inc., New York

13. Theophilus (1979) On divers arts. (Translated by Hawthorne JG and Smith CS). Dover Publications, Inc., New York
14. Doerner M (1984) The materials of the artist and their use in painting (revised edn.). Harcourt-Brace, San Diego
15. Romanian manuscript 1555, State Archives, Bucharest
16. Romanian manuscript 1825, Library of the Romanian Academy, Bucharest
17. Romanian manuscript 1251, Library of the Romanian Academy, Bucharest
18. Romanian manuscript 2151, Library of the Romanian Academy, Bucharest
19. Sandulescu-Verna C (1979) Erminia picturii bizantine. Ed. Mitropoliei Banatului, Timisoara
20. Romanian manuscript 5769, Library of the Romanian Academy, Bucharest
21. Gottsegen MD (1993) The painter's handbook. Watson Guptill Publ., New York
22. Romanian manuscript 4206, Library of the Romanian Academy, Bucharest
23. Romanian manuscript 2501, Library of the Romanian Academy, Bucharest
24. Romanian manuscript 1808, Library of the Romanian Academy, Bucharest
25. Kiplik DI (1954) Materialele colorate ale picturii, Editura de Stat pentru Literatura si Arta. Bucharest
26. Mihalcu M, Leonida MD, Maier RO (1990) Observations concerning preparative processes for materials used in medieval Romanian workshops. Revista de Istorie (Bucharest) I(4):386
27. Mayer R (1991) The artist 's handbook, 5th edn. Penguin Books USA, New York
28. Vitruvius on architecture (1931–1934) (editor and translated Granger, F.), G. P. Putnam's Sons, New York
29. Vasari G (1960) Lives of painters, sculptors and architects. Dover Publications, New York

Chapter 5
Adhesives Prepared from Vegetal and Animal Sources Used by Iconographers

Glues prepared from natural materials of vegetal and animal origin were used by iconographers in different applications: primer[1] recipes for painting on wood with tempera paints, on textiles, on dry mortar, in oil paints to make them shine, in adhesive primers used for gilding in manuscripts, in inks, and in some restorations and impregnations. All these adhesive preparations were similar to those used in Western Europe and in the Middle as well as the Far East.

Glues made from bones were not very common in Southeastern Europe. In the painter's manuals glue made from sturgeon bones is seldom mentioned and that prepared from animal bones almost never. The technical tradition had recorded that the best glues were those made from animal hide. Bovine hides were recommended most of the time, ovine ones less often. Buffalo hides were considered the best [1–4]. For this material to be less expensive, the artisans were encouraged by the *hermeneias* to use leftovers from hides skinned off the legs or even ears [2, 3]. The bovine and ovine hides used to extract glue could be either tanned or raw [4, 5].

Raw hide tailings were processed according to a technology known and practiced for a long time in the entire European area. Quicklime (burnt lime), obtained by roasting calcium carbonate from natural deposits, was slaked (hydrated) in a wooden trough [3, 6]. Following vigorous stirring a suprasaturated solution of calcium hydroxide was formed ("lime milk", slaked lime). After bringing the suspension to room temperature, hide tailings were soaked in it for about a week [1, 3, 4] until the tailings appeared at the surface of the suspension, marking the end of the process. Periodically during this time attempts were made to remove the pores from the hide. In a variant of this process it was recommended to boil the hides in water before tanning [1, 3].

In the next step the hide fragments were cleaned by scraping them with a special tool to remove leftovers (tallow, meat, dirt) which would pollute the extracted glue. When soaking in water was recommended, the process started in warm water followed by boiling for a short time [1, 3, 4]. Following this treatment the subsequent scraping was easier to perform and more efficient.

[1] Primer (undercoat) is a preparatory coating applied before painting. It affords better adhesion and durability of the paint to/on the surface, and provides additional protection for the material being painted. It is even more important when painting is done on porous materials (wood).

M. D. Leonida, *The Materials and Craft of Early Iconographers*,
SpringerBriefs in Materials, DOI 10.1007/978-3-319-04828-4_5, © The Author(s) 2014

To facilitate glue extraction, in the next step, hide tailings were chopped with a hatchet or an adze, on a wood board [2]. The fragmentation had to be incomplete, leaving pieces attached. The ensuing boiling in clean water was continued until the resulting solution of glue attained the targeted concentration. The manuscripts specify that concentration had to be allowed, but without "freezing" (gel was not allowed to form). While still warm, the liquid was strained through either a fine sieve or a clean cloth [3].

The next phase of the preparation, the same as for the process using tanned or raw hides and over the entire geographical area, was the further concentration of the glue solution in copper vessels to a point just short of gel formation [7]. Another *hermeneia* recommended to continue the heating/concentration of the solution "until the glue curdles" (thickens). During boiling continuous stirring was done using a wood rod. After completion, the glue solution was poured into metallic trays (thin layer) to cool. After the gel was formed, the leaf was taken out, cut into squares/rectangles and allowed to air-dry (24–48 h) on wood boards. Periodically these pieces were turned to allow uniform drying. In the last step these fragments were strung on a rope and dried "in the wind" [1], protected from solar radiation.

Concerning how to conduct the process, the Romanian *hermeneias* and the one compiled by Dionysius of Fourna are in agreement. The extraction had to be conducted under supervision because abundant foaming can occur and, consequently, overflow and loss of material. The Romanian manuscripts recommend to prevent foaming either by blowing air with a reed, or by adding a small amount of clean, cold water, or by removing the vessel in which the extraction was conducted from the heat source [3].

Concerning the time of the year when the extraction of glue from hides was to be done, the handbooks recommended to avoid the warm season, and solar radiation in general. The explanation given for this was "otherwise it spoils"[2, 4] or in another manuscript "not to stink" [1]. Today we would explain this by "to avoid contamination by microorganisms and fungi".

When such a glue was used in primer recipes several *hermeneias* [2, 3, 7, 8] recommended to use a fresh glue because an aged one loses its adhesive power. Prior to use, the glue was chopped and allowed to soak in water for about 12 h [3]. Furthermore, when the primer was used for an *iconostasis,* some technical manuscripts recommend that the primer be prepared immediately before use [3]. When used in a shiny paint, some *hermeneias* recommend to mix the glue with "strong (concentrated) ash lye" and beeswax.

The glue prepared from fish bones from the sturgeon family or from bones of a fresh water fish *Vimba vimba*[2] was used to protect metal work on religious objects[3] [1, 9], especially silver. This metal tarnishes easily under the action of sulfurous gases, due to silver sulfide (Ag_2S) formation. This adhesive material has been

[2] European fresh water fish found in rivers draining to the Caspian Sea, Black Sea, Baltic Sea, even the North Sea. It can be found as far east as Armenia and Azerbajdjan. It is also known as the vimba bream.

[3] Silver or gold were used to cover sections of an icon, book covers, reliquaries, or other objects. It was done to protect and it could be partial or over the entire surface. The metal sheet was either left as such or decorated by punching, hammering or etching.

known and used in Europe since the early Middle Ages. It had lower adherence and resistance compared to the glue made from bovine hides. It had the advantage that it could be used in applications where warm solutions of hide glue could not be used. This type of glue was used in some restorations [10], in filler mixtures (together with sawdust), and to glue together fragments of paper or wood [9]. Mixed with myrrh and alcoholic solutions the sturgeon glue was used by iconographers to prepare low quality varnishes (Chap. 6). Before use it was chopped with the handle of a knife and was boiled in water "until it melts" [1]. It was applied warm, in thin layers, after it had been strained through a threadbare cotton cloth. Some fish glue recipes included a colorant too ("good saffron") and myrrh.

A less usual adhesive used by the iconographers of Southeastern Europe was the secretions (slaver) of a gastropod (snail). Snail slime production was stimulated by introducing the flame of a candle in the opening of the shell (see Sect. 4.5). It was collected in a shell or another small container and used, mixed with gum arabic and gold powder, to write initial letters in manuscripts [1, 3, 4].

Mentioned and used since the eleventh century in Western Europe, the "cheese and lime glue" [11] was known and used in Southeastern Europe as well. The preparative technology in this region is slightly different from that described by Theophilus Presbyter [12]. A mixture of pot cheese and slaked lime[4] was heated (high flame) in a metallic "pan" and stirred using a wooden rod until it "looks like threads" [9]. This glue was mostly used in restorations and as binder in *gesso*[5] preparations for painting on wood.

Garlic juice was another adhesive of vegetal origin (*ilinocopia*) used by iconographers. Its extraction and uses were described in Sect. 4.4. Besides its use in manuscripts (in "gold ink" or adhesive for gold leaf), mixed with a black pigment it was used when a pattern was copied.

Another glue from a vegetal source was cherry glue [13], used in black ink recipes. It was used dissolved in clean water without further processing.

Gum arabic, an adhesive material extracted from exotic acacia trees (growing in Africa, Asia, and Australia) was imported, most of the time from North Africa. It was used dissolved in clean, warm water for gilding (see Sect. 4.5) and in ink recipes (Chap. 7). In ferro-gallic inks [14] and in those containing sugar or rock candy [15] it functioned as a binder. The same was the case for red inks where it was sometimes mixed with rock candy.

Egg white mixed with quicklime (CaO) was used to glue pieces of ceramic and glass. After the mixture was applied to both fragments, they were stuck together by binding around a rope and allowed to dry [16].

Potato glue was prepared as follows: the potato was peeled, washed, grated, allowed to macerate for approximately 12 h, squeezed by hand and then the juice was strained through a threadbare cloth. The thick product was boiled "well". This type of glue was prepared shortly before use [13].

---

[4] Casein, a milk protein, is soluble only in the presence of a base.

[5] White liquid made from chalk mixed with a glue solution or with casein. It is used as ground for painting on wood or other hard supports and in gilding.

References

1. Romanian manuscript (1795) Library of the Romanian Academy, Bucharest
2. Romanian manuscript (1555) State Archives, Bucharest
3. Sandulescu-Verna C (1979) Erminia picturii byzantine. Ed. Mitropoliei Banatului, Timisoara
4. Romanian manuscript 5769, Library of the Romanian Academy, Bucharest
5. Mihalcu M (1984) Romanian medieval values. Ed. Sport-Turism, Bucharest
6. Mihalcu M, Leonida MD (1980) Lime and technologies connected to its preparation and uses found in the recommendations of the *hermeneias* and in the Romanian mural painting. Rev Muzeelor şi Monumentelor Istorice şi de Artă XLIX(1):61–65
7. Dionysius of Fourna (1974) The painter's manual (Translated by Paul Hetherington). St. Vladimir Seminary Press, Yonkers
8. Romanian manuscript 3719, Library of the Romanian Academy, Bucharest
9. Romanian manuscript 4206, Library of the Romanian Academy, Bucharest
10. Romanian manuscript (1832) Library of the Romanian Academy, Bucharest
11. Cennini CA (1960) The craftsman's handbook (Translated by Daniel Thompson). Dover Publ. Inc., New York, p 68
12. Theophilus Presbyter (1979) On divers arts (Translated by Hawthorne JG and Smith CS). Dover Publications, Inc., New York
13. Romanian manuscript 1632, Library of the Romanian Academy, Bucharest
14. Romanian manuscript 3586, Library of the Romanian Academy, Bucharest
15. Romanian manuscript 1457, Library of the Romanian Academy, Bucharest
16. Romanian manuscript 2151, Library of the Romanian Academy, Bucharest

Chapter 6
Varnishes

The first varnish recipe known is one found in the manuscript of Theophilus Presbyter [1] and it is that of an oil varnish. A varnish is a solution of a resin in a volatile solvent. Brushed on a painted surface, it dries leaving a glossy, transparent, protective film. There are two types of varnishes: simple solution varnish (resin is dissolved directly in the solvent) and oil varnish (resin is melted together with a drying oil and a drier, then thinned with a solvent). From the outset, passing through Cenninni's Handbook [2], the De Mayerne manuscript [3], the Athonite, and the Romanian *hermeneias*, varnishes were always an object of study for those who studied materials employed by painters.

Varnishes, after application and drying to form a film, had a protective role, brightened the colors, and maintained the chromatic scheme. At the same time they strengthen the painted layer increasing its resistance to mechanic shocks, friction, and other accidental mechanical, physical or chemical strains [4].

This chapter focuses on varnishes based on linseed oil. They are oil varnishes, specifically solutions of natural resins in linseed oil. Sometimes the recipe included a solvent (such as *neft*[1]) as well. While linseed oil did not evaporate during "drying" and film formation (it underwent polymerization), the *neft* was totally removed, through evaporation. Due to changes occuring during "drying" after application on the painted layer, the film formed could not be easily dissolved, even when the same solvent was used (*neft* or linseed oil). These varnishes had an excellent durability when the icons were kept indoors and when the painted layer was on a flexible support (e.g. textiles) or on wood. When instead of linseed oil siccativated in the sun cooked linseed oil was used, the resulting film had a more pronounced darkening tendency and increased fragility.

Recipes for oil varnishes are less numerous in the handbooks of the Southeastern European iconographers (circa 18 %) than the ethilic ("brandy") ones (circa 63 %), the ones containing turpentine (circa 48 %), or *neft* (circa 21 %). In a varnish recipe linseed oil had the role of a solvent, vehicle, and binder. In Western Europe this type of varnishes was more common. In three quarters of the oil varnishes used

[1] Name used in the Romanian *hermeneias* sometimes for turpentine, sometimes for a mixture of solvents obtained from a rudimentary distillation of coal tar, petroleum, or peat. In Greek: *naphtha*.

M. D. Leonida, *The Materials and Craft of Early Iconographers*, SpringerBriefs in Materials, DOI 10.1007/978-3-319-04828-4_6, © The Author(s) 2014

by iconographers linseed oil is the only solvent. In almost 38 % of them, it appears together with *neft*. The latter was introduced in order to shorten the "drying" time of the varnish and to make it less thick and easier to apply (Table 6.1).

The raw materials used for varnish preparations were obtained either from local sources (linseed oil, autochthonous colophony[2]/*pegula*, chemical siccatives like $ZnSO_4$) or from import (*sandarac*, aloe, *santalon*, fine colophony/"the good colophony").

Linseed oil was obtained from clean flax seeds, without other oleaginous seeds present, by cold pressing, according to medieval European tradition. Sometimes the oil was siccativated prior to using it in a recipe, usually by prolonged exposure to the sun. Less often and later (19th century) siccativation was achieved at the time of the application by using chemical siccatives in the recipe (see Table 6.1). The number of such recipes is small, indicating iconographers' preference for linseed oil siccativated using a less violent procedure. There were also recipes using non-siccativated linseed oil [4–9].

In medieval Italy, besides linseed oil, hemp oil, poppy seed oil, and walnut oil were used as well. They were first exposed to solar radiation, in shallow trays, until a thin film of polymerized oil was formed at the surface. After removing the film, the oil left had a honey-like viscosity and was called "fatty oil". Varnishes prepared with boiled poppy seed oil, thickened in the sun or with a siccative, dried more difficultly and were inferior to varnishes based on linseed oil.

The quality of the linseed oil prepared in the workshops of the early Southeastern European iconographers was superior to that of the present day product. This is due to the high speed of the modern process (using siccative agents), more energetic and difficult to control. While the *hermeneias*, like manuscripts from all over Europe, mention lead compounds (carbonate, oxides) used as siccatives, present day practice uses lead, manganese, and cobalt oleates and resinates for the same purpose.

Rosin (colophony, *pegula*) from autochthonous sources was produced, in workshops from Southeastern Europe, in a brass vessel filled to one third only [5] with resin from local conifer species. The resin was heated until it became clear and no froth was produced anymore. Frothing was controlled either by removing the heating source or by blowing air [12]. Finally, the melted resin was poured into a vessel with cold water. *Pegula* was obtained as pieces (of variable sizes) of transparent material, with a light yellow tint [6].

Imported rosin ("the good colophony", "white colophony") was often used in varnishes together with another resin (see Table 6.1, [4–6, 9, 11]).

The use of rosin in varnishes was abandoned at a point, in spite of the glossy finish it imparted, due to its rapid tendency to darken, directly proportional with the intensity of the light the painted layer is exposed to. Another property of rosin is that it softens lightly upon heating (yielding a slightly sticky surface) and it hardens again upon cooling. This apparent shortcoming is actualy a desirable quality in a varnish. If it were dry all the time, it would not react to changes in temperature

[2] Old name of the resin rosin, still used in French and German. In Romanian: *pegula*

Table 6.1 Composition (%) of linseed oil-based varnishes used by iconographers in Southeastern Europe

Varnish	Non-siccativated oil	Siccativated oil	Rosin (autochthonous)	Rosin (import)	Sandarac	Aloe	*Santalon*	Neft	Minium	Litharge	Zinc sulfate	Reference
With santalon	–	33.33	–	–	–	–	66.67	–	–	–	–	[5]
With santalon	71.43	–	–	–	–	–	28.57	–	–	–	–	[6, 7]
Oil varnish	–	50	50	–	–	–	–	n.s.				[8]
With bezir[a]	–	57.14	42.86	–	–	–	–	n.s.	–	–	–	[10]
With santalon	–	66.67	–	–	–	–	33.33	–	–	–	–	[10]
Ol varnish	–	48.79	51.21	–	–	–	–	–	–	–	–	[7]
Oil varnish	–	12.42	87.58	–	–	–	–	n.s.	–	–	–	[5]
Oil varnish	–	57.14	14.29	28.57	–	–	–	n.s.	–	–	–	[5]
"Yellow varnish"	–	62.5	–	–	25.0	12.50	–	n.s.	–	–	–	[5, 6]
With santalon	–	64.21	–	33.16	–	–	12.63	–	–	–	–	[5, 6, 11]
oil varnish	61.49	–	–	11.11	–	–	27.40	–	–	–	–	[11]
"yellow varnish"	–	52.5	–	–	25	22.5	–	n.s.	–	–	–	[7, 10]
Oil varnish	–	55.86	22.35	13.41	–	–	–	8.38	–	–	–	[4]
"Yellow varnish"	5.40	–	–	–	27.05	13.55	–	54.0	–	–	–	[4, 9]
Varnish	–	37.04	37.04	7.40	–	–	–	18.52	–	–	–	[4]
"Oil shine"	72.92	–	–	–	–	–	–	–	18.24	5.84	3.0	[12]
Oil varnish	–	11.36	80.11	5.68	2.85	–	–	n.s.	–	–	–	[6]
Varnish	–	54.04	27.03	13.51	–	–	–	5.41	–	–	–	[11]
Varnish	–	55.55	22.22	13.89	–	–	–	8.33	–	–	–	[11]
Varnish	–	42.55	42.55	8.52	–	–	–	6.38	–	–	–	[11]
Oil varnish	–	55.86	22.35	13.41	–	–	–	8.38	–	–	–	[9]
"Yellow varnish"	–	2.78	–	–	27.77	13.89	–	55.56	–	–	–	[9]
Varnish	–	37.03	37.03	7.42	–	–	–	18.52	–	–	–	[9]

n.s.—not specified

[a] Vegetal oil obtained by the pressing, at ambient temperature, seeds or fruit. It was used as vehicle and binder in dyes, to impregnate/conserve wood, and even in foods

which affect the painted layer, triggering cracking which may affect the underlying painted layer as well.

Sandarac, a hard resin extracted from *Thuja occidentalis, Tetraclinis articulata, Cypress* conifers, and *Calitris quadrivalvis*, was imported. As the European West, iconographers in Southeastern Europe used it in hard and glassy oil varnishes, which were transparent, with a light yellow tint. In the oil varnish recipes used in this geographical area and which contained *sandarac*, very seldom a plasticizer (such as camphor) was used in spite of the resin's hardness, as was the case when other solvents were used.

This resin has been known and used since Ancient Rome. The De Mayerne manuscript mentions that this resin is what appears under the name amber in some sources. This latter name seems to be a generic one, due to the resin's color, "sometimes reddish, sometimes very pale". When the source of this resin was India, the color was very pale. The same manuscript [3] recommended to melt the resin in a silver spoon to get rid of the "dirty skin" and only then to dissolve it in an appropiate solvent. Due to the particular hardness of this material, Western manuscripts recommended to use it in mixtures with other (softer) resins. The De Mayerne manuscript contains the first known detailed discussion concerning the reversibility of varnishing. Another West European source [2] recommends use of *sandarac* in oil varnishes.

The aloe resin was obtained from imports as well. It was extracted from plants from the *Liliaceae* family (*Aloe africana, Aloe vera/vulgaris*). Oil varnishes containing it also contained *sandarac* and sometimes *neft* (Table 6.1, [4–6, 9, 10]. They were used sometime to imitate gold when applied on sections of icons covered in silver. In the Romanian area this resin was used in popular medicine as well.

Santalon was a hard resin extracted from the wood of several tropical trees from the species *Santalum,* the best known of which is the Indian sandalwood tree. In Dionysius of Fourna's *hermeneia* [5] this resin is mentioned under the name "sugar of *santalon*".

In Western Europe medieval artists used another varnish too, made with the resin *mastic*, a secretion of the plant *Pistachia lentiscus* which grows around the Mediterranean Sea. They were using it as a solution in cooked linseed oil in the presence of lead carbonate as siccative. This varnish (known as *meglip*) was not used by Southeastern European iconographers.

Dammar[3] was discovered later and is the last natural resin still used in varnishes (the rest are synthetic ones). It is a resin obtained from trees of the family *Dipterocarpaceae* growing in Indonesia and Malaysia. It has the advantage that it dissolves in white spirit but yields, after drying, a fragile layer. It is used only in the presence of plasticizers. There are no varnish recipes including *dammar* in the technical part of the *hermeneias*.

Oil varnishes were prepared in a very clean metallic vessel [10]. After breaking it into small pieces, the resin was melted (at temperatures higher than 300 °C) and the volatile fractions evaporated. After heating was stopped, the preheated oil was added [9]. Heating was resumed, accompanied by stirring with an iron rod (kept

[3] *Damar*—in references for artists' materials; *dammar* in Europe and in scientific literature.

red hot) until the resinous components were fully incorporated into the oil phase. During this latter step other recipe components were added (if needed). Adding a volatile solvent (such as *neft*) was done after the heating source was removed, to avoid fire hazards. A Western European manuscript [1] recommended to continue heating the solution of the resin in oil, in the presence of a lead-based siccative, until the volume was reduced to two thirds.

After having incorporated all the recipe components, in most cases, the varnish was filtered through a threadbare cloth. Subsequently it was kept in glazed ceramic containers or in well-stoppered flasks. Sealing the containers prevented evaporation of some volatile components, avoided undesired oxidation reactions, and prevented mechanical contamination with airborne particles.

The painter's handbooks recommended to apply the varnish "in the sun". Also, the first layer of varnish needed to be less viscous (in order to penetrate more deeply the painted layer, "not to be runny"). The anonymous compiler of a manuscript [7] added: "if it curdles (too viscous) put a little *neft* to stretch it, and if you do not have *neft,* put some raw (uncooked) oil".

In Western Europe varnishing was done "in the sun" too but Theophilus Presbyter recommends to preheat the painted layer first, then to apply the varnish with a brush and then to spread it well by hand.

Table 6.1 summarizes some regional characteristics concerning varnishes. Approximately 30 % of the recipes contain two components. Among them, the highest number (~85 %) include linseed oil siccativated by solar radiation and only ~15 % recommend uncooked linseed oil. Rosin from local sources is recommended in 57 % of the varnishes, imported rosin in 9 %, and 38 % of the varnishes use mixtures of the two types of rosin. *Sandarac* mixed with a softer resin is listed in 31 % of the recipes while 22 % contain a mixture of *santalon* and imported rosin. *Neft,* the second solvent, appears in a specified amount in 38 % of the varnishes. Only one varnish recipe contains siccatives (natural compounds of lead and zinc). The "yellow varnishes" were recommended for use on silver-covered sections to "make them yellow" (imitate gold) [12].

As we previously mentioned, starting sometime around 1830, *dammar* replaced *sandarac* and then the time of synthetic resins came: first the cellulosic ones, then polyvinyl acetate, polymetacrylic esters, polycyclohexanone with reduced molecular mass, etc. Siccatives and plasticizers diversified too. Behind all these inventions the drive remained the same: how to impart radiance to a painting, brightness, resistance to mechanical shocks and to ageing, without interfering with the painted layer and with the artisan's intention.

References

1. Theophilus Presbyter (1979) On divers arts (Translated by Hawthorne JG and Smith CS). Dover Publications, Inc., New York
2. Cennini CA (1960) The craftsman's handbook (Translated by Daniel Thompson). Dover Publications, Inc., New York

3. De Mayerne TT (2001) Lost secrets of Flemish painting including the first complete English translation of the De Mayerne Manuscript, B.M. Sloane 2052 (1st ed.), Donald C. Fels (commentary and drawings). Alchemist, Hillsville
4. Mihalcu M, Leonida MD (1987) Varnishes and lacquers in medieval Romanian art. The annual conference of the history and archaeology Institute A.D, Xenopol, Iasi, 9–10 May
5. Dionysius of Fourna (1974) The painter's manual (Translated by Paul Hetherington). St. Vladimir Seminary Press, Yonkers
6. Grecu V (1936) Carti de picture bisericeasca bizantina. Ed. Glasul Bucovinei, Cernauti
7. Romanian manuscript 1555, State Archives, Bucharest
8. Mihalcu M, Leonida MD (2009) Din tainele iconarilor romani de altadata. Ed. Eikon, Cluj-Napoca
9. Sandulescu-Verna C (1979) Erminia picturii bizantine. Ed. Mitropoliei Banatului, Timisoara
10. Romanian manuscript 2151, Library of the Romanian Academy, Bucharest
11. Mihalcu M, Leonida MD, Maier RO (1990) Observations concerning some preparative technologies used in medieval Romanian workshops. Rev Istorica 1:86–92
12. Romanian manuscript 1795, Library of the Romanian Academy, Bucharest

Chapter 7
Inks Prepared and Used by Iconographers

During the Middle Ages writing was done on varied materials (parchment, paper, wood, wood tablets covered in wax, glass, stone, mortar) and using varied tools (quills, nibs, brushes, by incision, by carving). The observations included in this chapter are concerned with inks used to write on paper and parchment only. They were used for official documents, but also to transcribe the painters' handbook and to write on manuscripts (some of them illuminated[1]). Some originated in princely, judicial or ecclesiastical offices. There are also numerous manuscripts written by clergymen or laymen who exchanged letters, translated or copied other manuscripts, or recorded events, balances, thoughts, recipes of interest for everyday life or for practicing different trades. Knowing the composition of the inks, the way to prepare and to use them, is of interest for the history of science, for those who conserve and restore art objects, as well as for their custodians.

The vast majority of medieval official documents from Southeastern Europe which were preserved were written in black ink, although there are a number written in blue [1]. However there are numerous cases when only a part of a manuscript was written in ink of another color than black. In some, titles (of the manuscript, of chapters and subchapters), the first letter of a chapter or paragraph, some fragments of the text, some annotations, or the titles and signatures of officials were written in red ink. In general, red was used to draw attention to that section of the text or, in the case of an official document, to make corrections to the initial provisions which had been subsequently reconsidered.

The inks were prepared by the future users or other specialized artisans using recipes handed down over generations or collected from foreign sources and adapted to the local conditions and availability [2]. The importance of inks is shown by the high number of recipes found in *hermeneias,* in *miscellanea*, in calendars, and on the margins of some manuscript pages [3–17]. The Athonite *hermeneias* were among the sources of these recipes in a time when Byzantium and what followed it was a model for the spiritual life in Southeastern Europe. However, while there are only a couple of such recipes in Dionysius of Fourna's well-known *hermeneia* [18],

[1] Manuscripts in which there are colored decorations (first letters, miniatures, borders). At the beginning the term was used only for manuscripts decorated with gold and silver. Today any manuscript from Europe or the Middle East having colored decorations is considered as such.

M. D. Leonida, *The Materials and Craft of Early Iconographers*, SpringerBriefs in Materials, DOI 10.1007/978-3-319-04828-4_7, © The Author(s) 2014

in the Romanian manuscripts the number is much higher, their compositions and their preparative technologies much more complex and, often, presented with more detail. Some of these recipes were known previously in the European area, but in most cases they show interesting innovations and local adaptations. Others—those obtained from fruit or leaf juice, sap from plant stalks or roots and those prepared from infusions of the same materials—can be obtained in easy to control conditions, using simple recipes which can be followed by anybody. They fully show the technical ability and the inventiveness of the artisans as well as a deep knowledge of the local flora and of the popular chromatic preferences.

Information concerning the composition of the inks used by medieval artisans, and the technologies used for their preparation, can be obtained either from studying the recipes found in the manuscripts mentioned above or by physico-chemical analyses. Some types of analyses have the drawback that they are destructive and require elaborate sample preparations (TLC, DESI-MS[2] [19]), while others (FT-IR-ATR[3]) are rapid, non-destructive and do not require any sample preparation [20].

A perusal of the ink recipes found in the *hermeneias* shows concern for achieving certain qualities in the product. A color as intense as possible was desirable and, sometimes, that was achieved later, after the writing had been done. Another quality interesting for an ink was its fluidity which would guarantee fine lines and easy flow off the nib. The adherence of the ink to the paper (without penetrating too deeply and without forming a halo around the written lines) and the resistance to the physical, chemical and biodegrading agents were qualities also targeted by those who created ink recipes. A succinct presentation of some ink found in the painter's handbooks of past centuries will show not only these concerns but the successful outcomes as well.

7.1 Black Inks

Recipes for black inks based on iron compounds of tannic acid (iron gall inks) were the most common. Greek manuscripts from the fourth to third centuries B.C. showed that the same was true in that area at that time. It is not clear when this type of ink started to be used in other parts of Southeastern Europe.

The raw materials used in their preparation were: oak apples (gallnuts, a natural source of tannic compounds, Fig. 7.1), ferrous sulfate (*copperas*, provider of the sulfate anion and, more importantly, of the ferrous cation), gum arabic and some autochthonous saps (not only for the adherence of the ink to the paper but also to adjust the density in order to stabilize the suspension of the colored tannic compounds). The most frequently used solvent was white wine, but mixtures of wine

[2] Desorption electrospray ionization mass spectrometry—atmospheric pressure ion source which ionizes inks (in this case) or other liquids, gases, even solids to be analyzed by mass spectrometry.

[3] Fourier transform infrared attenuated total reflectance is a sampling technique using the property of total internal reflection that is coupled with infrared spectroscopy.

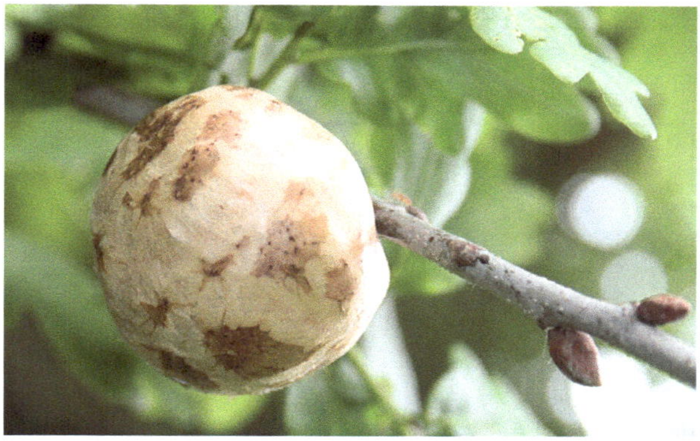

Fig. 7.1 Oak apple/gallnut. (http://en.wikipedia.org/wiki/File:Oak_apple.jpg)

and water, water, and mixtures of vinegar and water were reported as well. In some recipes alum[4] appears too (to counter the effect of hard water) and, very seldom, sodium chloride (table salt, as preservative).

Table 7.1 shows recipes for several black permanent inks found in Romanian *hermeneias*. Concerning the raw materials recommended, two approaches were followed. In the first, a ferrous salt (*copperas*) was reacted with organic vegetal extracts (containing tannic compounds). The extracts, containing mucilaginous compounds, were obtained from oak gallnuts, in boiling water. In the second variant, the same solution (less rich in tannic compounds) was mixed with an extract (in boiling water only and without any preservative) from crab apple skin (rich in pectin).

The oak gallnuts, excrescences which appear on the back of oak leaves following the sting and deposition of fertilized eggs by a wasp from the *Cynipidae* family (due to the irritants released by the larvae), were the raw material providing the tannins (50–75 %). They were harvested before the young generation of insects came out, at the time when the content of tannic compounds was at a maximum. The gallnuts were not supposed to be light and hollow, they needed to show a compact mass (in section) without perforations bored by the larvae developed inside. The size of the gallnuts was smaller than a cherry. After being harvested, they were dried and then ground in rudimentary mortars.

Although there are many vegetal compounds containing tannic compounds (bark, leaves, or wood of conifers, willow, oak, poplar), only the oak gallnuts, ash tree bark and crab apple skins are mentioned as sources of tannins in the preparation of black inks in the *hermeneias*. The crab apple skins are not mentioned in the Athonite manuscripts. The extensive use of oak gallnuts makes sense since they are the richest and purest sources of tannic acid compounds. There is no mention in the handbooks of the use of oak bark leftovers from tanneries, as was often found in other European areas.

---

[4] As a chemical compound it is the hydrated potassium aluminum sulfate (potassium alum) with the formula $KAl(SO_4)_2 \cdot 12H_2O$. The name is also used for double sulfates of a monovalent cation (alkali or ammonium) and a trivalent one (aluminum or chromium).

Table 7.1 Black ink recipes based on tannic compounds

References	Oak gallnut (%)	*Copperas* (%)	Gum arabic (%)	Cherry tree sap (%)	*Alum* (%)	Sodium chloride (%)	White wine (%)	Vinegar (%)	Water (%)
[2]	8.35	4.17	3.55	–	0.47	–	41.73	–	41.73
[2]	6.63	4.42	4.42	–	0.43	1.29	–	55.21	27.60
[21]	31.30	8.14	1.88	–	–	–	58.68	–	–
[21]	8.22	4.12	–	4.12	–	–	83.54	–	–
[22]	4.32	2.88	2.88	–	–	–	–	–	89.92
[18]	9.56	34.13	34.13	–	–	–	22.18	–	–
[23]	9.56	34.13	34.13	–	–	–	22.18	–	–
[24]	9.56	34.13	34.13	–	–	–	22.18	–	–
[1]	9.56	34.13	34.13	–	–	–	22.18	–	–
[25]	4.02	6.02	5.62	–	–	–	–	–	84.33
[4]	4.02	6.02	5.62	–	–	–	–	–	84.33
[1]	4.02	6.02	5.62	–	–	–	–	–	84.33
[3]	4.02	6.02	5.62	–	–	–	–	–	84.33
[6]	6.88	4.28	2.14	–	–	–	86.80	–	–

The ferrous sulfate (*copperas*) was obtained either from natural deposits or from commercial sources. It was used because the more efficient ferric salt was not available at that time. Gum arabic was obtained from commercial sources, while cherry tree and plum tree sap were harvested from nature and used as such.

The technological steps included in the preparation of the first type of black ink were the same in the secretive *hermeneias* as in other manuscripts found in the European area. First, the solid components were brought by repeated grinding and sifting procedures (see Chap. 8) to a very fine granulation. This was done to facilitate the extraction process by enhancing the contact between the reactants. The resulting powders, with or without prior mixing, were introduced in a colorless, water-based solvent. The mixture was boiled, in order to facilitate the extraction in water of the natural tannic compounds and their subsequent reaction with the ferrous salt, and then cooled. The boiling also destroyed the possible fungi which would otherwise grow and spoil the ink [5–9, 18]. Some recipes did not require boiling. When required, the duration of the boiling was controlled by the repeated appearance (3–6 times, variable from one recipe to another) of strong frothing. Suggestively, this frothing is compared in the manuscripts to the one produced when Turkish coffee (a very popular beverage in that geographical area) is prepared (by prolonged simmering). When this happened, the heating was stopped and the liquid was allowed to settle, in order to avoid spillage and material loss.

When the recipe did not require boiling, the mixture was exposed to solar radiation for 4–7 days. During this time, in order to facilitate extraction, homogenization and dissolution of the mixture, the vessel was periodically and vigorously shaken.

Following either procedure (with or without boiling) the supernatant was separated by filtration through either a thick cloth (made of hemp) or a very fine sieve. Sometimes it was recommended [6, 7] for the resulting filtrate to be stored in bottles which were subsequently exposed to sunlight for 20 days. This was allowing for oxidation reactions to be completed and afforded an ink with a darker color at the time of its use. When crab apple skin was used, a cold extraction was done first (in 1:1.5 w/w water:skins) for approximately 7 days [1, 24, 25]. In some manuscripts sun light exposure is recommended for the duration of this extraction. The addition of the other components (oak gallnuts, *copperas,* vegetal sap) was done at a later time. The subsequent boiling was longer in this type of recipe, until the volume of the liquid mixture was halved. In some manuscripts [15, 26] it was recommended for the solvent to be added only after the final filtration.

The reaction between the ferrous salt and the tannic compounds, both variants mentioned above, was the only one of interest in the preparation of this type of inks because it resulted in dark color compounds. The reaction between *copperas* and the mucilaginous compounds was adversely affecting the quality of the ink by increasing the viscosity and the tendency to grow mold, as well as leading to discoloration and formation, in time, of a precipitate (result of a reaction between the tannic compounds and the mucilaginous matter). The ensuing filtration was reducing these risks and was of paramount importance for the stability of the resulting ink.

Following filtration, the density of the liquid was decreasing considerably thereby increasing the danger for the colored compounds of interest to precipitate and decrease the intensity of the ink color. Adding gum arabic or a vegetal sap had a stabilizing effect on the colored dispersion. Very seldom sugar was used to this end. During that time it was expensive, not easy to find, and it increased the risk of fungal and mold growth.

Examination of the data in Table 7.1 leads to interesting observations concerning the ratios between the components in different black inks. Concerning the proportion of water, we observe a large variation, from ~22 % in the process where the solvent was a mixture of wine and vinegar, up to ~89 % when only water was used.

A large variability is observed in the ratio of oak gallnuts used, always higher in the first type of process (6.6–31.3 %) than in the second (~4 %). In the latter case the extraction in boiling water of the crab apple skin was contributing tannic compounds to the mixture as well. This would lead, sometimes, to a darker color (result of a higher concentration of tannic compounds in the ink) for the inks obtained from the second technological process.

Concerning the proportion of ferrous sulfate used, several observations are in order. The recipes mentioned in the painter's handbooks can offer only an approximate idea about the amounts used. Much later than the period of interest methods determining exactly the concentration of tannic compounds in different natural sources became available. They made it possible to determine the exact amount of *copperas* required when a certain vegetal source is used. Due to the fact that a ferrous salt was used, not a ferric one (since it was impossible to find it at that time), the maximum intensity of the ink color was attained sometime after the writing was completed. The time corresponded to the amount of time required for the oxidation of the ferrous cation to be completed. This was true for the vast majority of the inks used at the time.

In all the recipes the amount of *copperas* exceeded the stoichiometric one. An excess of ferrous salt acted sometimes not only as a decomposing agent for the colored ferro-tannic salts of interest but as an agent decreasing the resistance of the ink to humidity (the excess of ferrous sulfate being water soluble). Due to this fact, the color of inks with such a flawed composition faded over time.

There are sections in older manuscripts written in black ink, where the ink appears to have a brownish color. This condition can be due to the conversion of part of the excess ferrous sulfate into a rust-colored basic compound, adduct of ferric sulfate and ferric oxide:

$$12\,FeSO_4 + 3\,O_2 \rightarrow 4\,Fe_2(SO_4)_3 + 2\,Fe_2O_3 \tag{7.1}$$

To assign all the drawbacks mentioned to lack of technical knowledge would be misleading. An excess of *copperas* was used because the centuries-long experience recorded that this salt is a good disinfectant and this was important at a time when not too many such substances were known. The increased proportion of ferrous sulfate was justified by its anti-mold effect too. Careful examination of the ink recipes found in the *hermeneias* shows that, although in all a ferrous sulfate excess is recommended, seldom this exceeded the limits beyond which the writing appeared rust-colored or the solubility became inadequate. Most of the time the proportion of ferrous salt was maintained within the limits for an effective antifungal effect. Furthermore, in the cases when an excess of *copperas* was used, the amount of dispersant-adhesive was accordingly increased.

Another category of black ink found in the painter's manuals is the one in which another material was used besides oak gallnuts, *copperas,* and cherry tree sap. This was a colored organic matter obtained by extraction in boiling water from the wood of an exotic tree *Haematoxylon campechianum*. This dye was used for textiles too and for Easter eggs as well. In inks its use was mentioned first during the nineteenth century. The preparative technology of this ink was basically the same. After boiling for ~2h the mixture of oak gallnuts, wood extract and water, ferrous sulfate and cherry tree sap were added to it [1, 4].

Some *hermeneias* mentioned another type of black ink made by extraction from ash bark (*Frasinus*). The Romanian manuscripts recommend an extraction time shorter by 50% compared to the Athonite ones. It was followed by the reduction of the volume of the solution to half by boiling while at Mount Athos only "brief boiling" was recommended.

Finally another black ink was the one in which the pigment was carbon black. This ink, obtained by suspending carbon black in water (using vegetal sap), had a much better resistance to aggressive agents and a more intense color. Concerning the carbon black, it was prepared as described in Chap. 3 (Sect. 3.4) and it was obtained as very fine particles which were used as such, without further grinding. The fine soot deposited on the bottom of buckets was harvested by vigorous hammering of the vessels it was deposited on.

It is likely that other black pigments obtained from vegetal sources were used as well. Among those mention has to be made of those obtained by the incomplete combustion of organic materials such as bones, oak wood, birch wood, grape vine, walnut shells, peach pits (see Sect. 3.4). The accurate identification of these

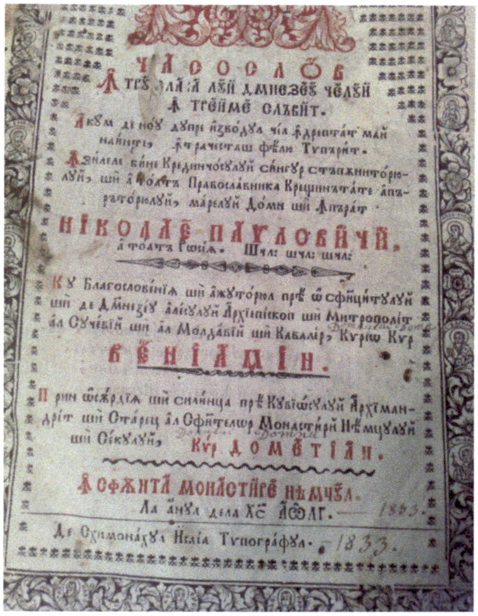

Fig. 7.2 First page in a Romanian religious book (*ceaslov*) where red and black inks were used. Photo: M.D. Leonida

materials—even in the conditions of today's modern laboratory—is difficult. All these inks were prepared in the same way, by suspending the pigment in brandy together with a small amount of plum tree sap [27].

7.2 Other Inks

Red inks were used before the twentieth century in official documents and title pages or first pages of books and chapters in books (Fig. 7.2)[5]. The most used pigment for red inks was cinnabar, due to its appealing qualities (vivid color and resistance to the action of external agents). This type of ink too was known since Antiquity, from Europe to the Far East. Several hues of red can be found in different manuscripts where it was used due to different concentrations of the pigment in the inks, to different purities, and to differences in storage conditions. It is known that dishonest merchants of times past were mixing cinnabar (difficult to find, expensive, and with lower density) with *minium* (easily accessible, much cheaper, and higher in density). When this happened, due to the latter's sensitivity to sulfurous gases, the hue of the ink darkened over time. Sometimes *litharge* was also substituted for cinnabar. During the stepwise preparation this was converted into *minium* too. Cinnabar was obtained from commercial sources, local deposits or prepared in the artisanal workshops, through "dry" synthesis, from mercury and sulfur (see Sect. 3.1).

[5] Religious book used in the Orthodox religious service mainly by cantors (not priests or deacons).

In the *hermeneias, miscellanea*, and other similar manuscripts, as well as in the annotations found on the margins therein, numerous recommendations concerning the preparation of red inks containing cinnabar can be found. The recipes contain all or only some of the following components: cinnabar (the pigment), gum arabic, rock candy or vegetal sap (the binder), wine, vinegar or water (the vehicle), myrrh, and brandy [5, 6, 8, 10, 11, 13, 14, 16, 26]. Sometimes in these inks a small amount of egg yolk emulsified in water was introduced [13, 14]. Differently from the black inks, in the recipes for the red inks the proportions of components are not specified in the painter's manuals.

The preparation of the red inks was starting with the grinding, in mortars, ("rubbing on a stone") and repeated sifting procedures of the cinnabar (see Chap. 8). The "rubbing on stone" was done in the presence of a very small amount of water. The other components were brought to the same fine granulation separately, by repeated grinding and sifting as well. The gum arabic and the rock candy were dissolved separately in a small volume of water. Subsequently all the ingredients were added to the solvent (vehicle) under vigorous and efficient stirring. The ink, filtered or not (depending on need), was then dispensed into inkpots and inkstands. Concerning what we call quality control nowadays, this was simply done by observing the intensity of the color, the fluidity, the shine, and the color the ink had after the writing was done and the ink had dried.

Wine, brandy and vinegar were used as vehicles for the inks. Without insisting on well-known technological details concerning them, we shall mention less known facts related to the vinegar and brandy used in the preparation of some inks in the geographical area of interest. Vinegar, prepared only from wine, was obtained through the same type of process as the alimentary one, only in more rigorously controlled conditions and of higher concentration. In the fermentation different additives were used (fermented dough, raisins, fried lentils, baker's yeast). The vessels used for this process were covered (to be protected from mechanical impurities) and kept in warmer places of the household. The control for the concentration of the alcohol (prepared in stills) was done putting gun powder and alcohol in a spoon and lighting the mixture. If the gun powder ignited and, after the flame went out, no liquid was left in the spoon, the alcohol was considered concentrated enough. This was actually a confirmation, very simply and resourcefully obtained, of the fact that the alcohol had attained the concentration of the ethylic azeotrope or higher [2, 18, 24].

Green inks were obtained from the sap of a plant (*Ruta graveolens*) and a solution of copper acetate (finely ground and "rubbed" with a little vinegar) [12, 14]. Sometimes, in order to obtain a warmer hue, a little saffron was introduced as well [8, 11, 12, 14, 15]. *Ruta* and saffron were obtained from nature and the sap was extracted in the household. Copper acetate was obtained in local workshops (see Sect. 3.2). The quality control for the green inks was done in the same way as for the black and red inks.

Blue ("purple") ink was obtained from the sap obtained by squeezing the fruits of the common elder (*Sambucus nigra*) to which alum and vinegar were added (a fourth of the amount of vegetal sap [12, 14, 15]) and, sometimes, tiny amounts of human urine [14, 15].

Other raw materials obtained from vegetal saps and used in the preparation of artisanal inks were alder (*Alnus glutinosa*) buds, green walnut shells, dogwood fruit. Sometimes vegetal glue, rock candy or even gum arabic was added to those inks.

Even gold ink was sometimes used in writing but much less [1]. A partial explanation of this situation is the fact that gold was used to write on parchment but, as we already mentioned, most of the manuscripts which reached us are written on paper.

A recipe for a golden ink used as pigment fine gold powder, snail slime (Sect. 4.5) as a binder, alum as conserving agent and gum arabic as adhesive [16, 17]. The alum and gold powder were "rubbed" first in a rudimentary stone mortar. After the desired granulation was attained, gum arabic was added too. This recipe is mentioned in the Dionysius of Fourna's *hermeneia* too [18]. Another ink of this type was prepared by mixing equal parts of a glue solution, ash lye wash and wax. The mixture was heated and well homogenized. Subsequently the gold powder was introduced under continuous stirring. After writing and drying, burnishing was done by using an animal (dog or wolf) tooth [27]. Sometimes first letters in a chapter or initials were done in gold leaf [28].

Concerning the use of silver ink in writing, it was done very seldom. There are still numerous recipes of this type of ink in the technical manuscripts addressed to iconographers [4, 8, 13, 15—17, 26, 27].

A few painter's handbooks also recommend methods to remove ink spots from textiles [5, 8, 26]. The stained piece of cloth was first steamed by keeping it above a pot with boiling water. The spot was then rubbed with lemon juice to convert the colored iron tannates into colorless, water soluble citrates, and table salt was sprinkled on it (absorbent). Finally the spotted cloth was well washed with water.

In another method [23] a solution was obtained by steeping and stirring (for 24 h) a mixture of ~600 g water, ~50 g ash from wine yeast, a "little" baking soda, and two lemons cut into small pieces. After 24 h, the liquid was filtered through a clean cloth. The ink stains were removed by wetting them and rubbing with the filtrate followed by washing with clean water.

7.3 "Disappearing" Inks

As a curiosity, we have to mention here the recipes given in the Southeastern European *hermeneias* for disappearing inks. When used, the writing appeared after treating the surface with a certain reagent. They are what we call today "invisible inks" [3, 23]. Their purpose was to transmit information meant to be seen only by a certain person and they were often used in times when wars and adversity were pretty common, or to transmit information addressed to exclusive groups, like that of the iconographers.

A classification could be according to the raw materials used to prepare them: (a) based on vegetal extracts used as such (onion, lemon [29, 30]); (b) chemical compounds obtained by extraction from plant sources (gallnuts/gall apples; see Sect. 7.1 [29]); (c) some beverages prepared in the household [30, 31].

Another criterion for the classification of invisible inks used by iconographers and other artisans is the type of process used to make writing appear. The physical one, used by far the more often, was heat. One handbook [31] recommends to "....use as much heat as needed....". Most of the invisible inks of this type were made of either lemon or onion juice [29–32].

Another type of "disappearing" inks used a chemical agent to make writing appear. This type of ink was made using an aqueous extract of oak gallnuts to which a little soap was added. The developer was a solution of *copperas* with which the written surfaces were wetted. The reaction was irreversible. The writing appeared black and could not be made invisible again.

After the beginning of the seventeenth century solutions of cobalt compounds started to be used. The writing done using them disappeared after drying. Following light heating, the writing appeared again. Depending on compound, after cooling, the writing would stay or disappear again.

The recommendations warn that a new quill or nib has to be used with such an ink for traces of ink from previous uses not to color the invisible ink. It is also recommended, for the cases when the oak gallnut extract is slightly colored, to dilute it with water.

References

1. Vartosu E (1968) Paleografie romana-chirilica. Editura Enciclopedica, Bucharest
2. Mihalcu M (1984) Romanian medieval values. Editura Sport-Turism, Bucharest
3. Romanian manuscript 98, Library of the Romanian Academy, Bucharest
4. Romanian manuscript 730, Library of the Romanian Academy, Bucharest
5. Romanian manuscript 1283, Library of the Romanian Academy, Bucharest
6. Romanian manuscript 1620, Library of the Romanian Academy, Bucharest
7. Romanian manuscript 1632, Library of the Romanian Academy, Bucharest
8. Romanian manuscript 1825, Library of the Romanian Academy, Bucharest
9. Romanian manuscript 2387, Library of the Romanian Academy, Bucharest
10. Romanian manuscript 3590, Library of the Romanian Academy, Bucharest
11. Romanian manuscript 3719, Library of the Romanian Academy, Bucharest
12. Romanian manuscript 4353, Library of the Romanian Academy, Bucharest
13. Romanian manuscript 5188, Library of the Romanian Academy, Bucharest
14. Romanian manuscript 5189, Library of the Romanian Academy, Bucharest
15. Romanian manuscript 5769, Library of the Romanian Academy, Bucharest
16. Romanian manuscript 5952, Library of the Romanian Academy, Bucharest
17. Romanian manuscript 1555, State Archives, Bucharest
18. Dionysius of Fourna (1974) The painter's manual. (trans: Hetherington P). St. Vladimir Seminary Press, Yonkers
19. Ifa DR, Gumaelius LM, Eberlin LS, Manicke NE, Cooks RG (2007) Forensic analysis of inks by imaging desorption electrospray ionization (DESI) mass spectrometry. Analyst 132:461–467
20. Feustel M, Briggs JL (2009) Rapid analysis of inks on paper by viewing FT-IR-ATR spectroscopy. Spectroscopy solutions for materials analysis, LC-GC special issue, August 1:1
21. Pamfile T (1910) Industria casnica la romani. Trecutul si starea ei de astazi. Tipografia Cooperativa, Bucharest
22. Pamfile T, Lupescu M (1914) Cromatica poporului roman. Socec & C. Sfetea, Bucharest
23. Vasilescu V (1968) Rev. Arhivelor (new series), 11, pp 253–258

24. Romanian manuscript (1904) Iconographia, arta de a zugravi biserici si icoane bisericesti (2nd edn of the manuscript edited in 1891 by Ghenadie of Ramnic and New Severin). Tipografia Cartilor Bisericesti, Bucharest
25. Stahl H, Bogdan PD (1936) Manual de paleografie slavo-romana. Fundatia pentru literature si arta, Regele Carol II, Bucharest
26. Grecu V (1936) Carti de pictura bisericeasca bizantina, (ed) Glasul Bucovinei, Cernauti
27. Lupu N (1977) Arhivele Olteniei, vol. VI:114
28. Mihalcu M, Leonida MD (2002) Adhesive materials used in gilding by Romanian medieval icon painters. Etudes et Documents Balkaniques et Mediterraneens 25:53–70
29. Romanian manuscript 2151, Library of the Romanian Academy, Bucharest
30. Romanian manuscript 2135, Library of the Romanian Academy, Bucharest
31. Romanian manuscript 1151, Library of the Romanian Academy, Bucharest
32. Mihalcu M, Leonida MD (2013) The hidden face of form and color (2nd edn). Editura Eikon, Cluj-Napoca

Chapter 8
About Grinding

Like the Western European artisans, the iconographers from Southeastern Europe were grinding their materials, according to the needs of different techniques used, in their own workshops. Practice and tradition imprinted into and transmitted through a remarkable collective memory the fact that particle size (especially that of pigments) was very important (for hue, amount consumed, shortening the working time) in the process of making the icons. To emphasize the importance of grinding, in the fourteenth century Cennino Cennini in his “The Craftman’s Handbook” [1] was saying how to do it for cinnabar: “…and grind it with clear water as much as you ever can; for if you were to grind it every day for 20 years, it would still be better and more perfect.” Similarly, Turquet de Mayerne[1] in his seventeenth century manuscript [2] describes how different colors have to be ground and in which medium (water, oil, etc).

The difference between East and West was the greater importance the former was attributing to apprenticeship and continued practice compared to information transmitted in writing on this topic. For instance, like their West European colleagues, the Southeastern artisans knew that *lampblack* is obtained as a very fine powder (Sect. 3.4), hence there is no recommendation in any of the painter’s handbooks on its grinding. However there are more recommendations concerning grinding in the Greek and Romanian *hermeneias* compared to the Russian technical manuscripts.

The steps to reach the desired fineness were slow and discontinuous, the breaks allowing uniform and full penetration of the liquid vehicle into the granules. The purpose of grinding (“rubbing”) the solids used by the iconographers in their work was not only to ease subsequent steps (synthesis or painting) but to obtain some interesting chromatic effects as well. When preparing adhesives used in gilding, by grinding the pigments, plasticizers, siccatives and other additives, a thorough mixing of the solid components of the recipe was targeted (Chap. 4). In varnishes (Chap. 6), grinding was used to enhance the dissolution of solid recipe components (resins, plasticizers), as well as in some inks (warm or cold solvent). For pigments,

[1] Sir Theodore Turquet de Mayerne – Swiss physician who treated kings of France and England. He performed chemical and physical experiments, prepared pigments and cosmetics, and introduced calomel to medical applications.

M. D. Leonida, *The Materials and Craft of Early Iconographers*,
SpringerBriefs in Materials, DOI 10.1007/978-3-319-04828-4_8, © The Author(s) 2014

grinding extended their covering ability and afforded superior optical characteristics and a better painted layer.

The iconographers had noticed that finer particles for recipe components were beneficial for reactants used in chemical syntheses performed in their own workshops (e.g. synthesis of cinnabar). They were also aware that too fine a cinnabar obtained from commercial sources makes adulteration with *minium* (higher density, cheaper, inferior hue, Sect. 3.1) impossible to detect by naked eye (the only method available at the time). It is clear in the painter's handbooks that they were aware that there is an optimum granulometry for each pigment and, at finer ones, some optical characteristics are lost and work is spent unnecessarily. The anonymous authors indicated, although not always in great detail, the desirable size for each pigment.

8.1 One-step vs. Several-Step Grinding

In general, the *hermeneias* recommended, in different ways, when it was necessary to grind a recipe component and if it has to be done separately, together with some or with all the other components. There are cases when there is no recommendation concerning grinding at all. For example, in a recipe for an alcoholic varnish [6] there is no reference to grinding the resins (*sandarac* and *mastic*[2]). The same situation can be found in the recipe of a ferro-gallic ink [7] wherein nothing was mentioned about the grinding of the solid components, a detail which, probably, was considered trivial.

Sometimes in the painter's handbooks it is recommended very clearly to grind all the recipe components together. In the recipe of another alcoholic varnish [8] with "many materials" it is required to grind together *sandarac, copal*[3] *(anime), elemi*[4], and camphor (plasticizer). Manuscript [9] recommends for a "spirit varnish" to grind well rosin (*colophony*) with *sandarac* and then mix them with alcohol and turpentine.

Grinding together the ingredients can be found for an adhesive mass used in gilding (*gulifarba* type) for which ochre, *verdigris,* minium, and *lime white* needed to be ground together [3]. In [8] recommendations about grinding are given for several types of adhesive preparations: for an *ambole* (Sect. 4.2) the solid recipe components (*bole/vole/vol,* ochre, minium, and paper ash) need to be ground together ("...put them together on a slab and grind them mighty well..."). In another adhesive ("God's *ambole*") for the solid components (*Tarigrad ochre*, bole, and iron trioxide) it is recommended to "...ground them well the mixings..." [8]. For yet another adhesive (a *poliment*) the same manuscript was asking for all the ingredients (*Tarigrad ochre,* minium, tallow, and paper ash) to be ground together. For the preparation of a *murdent* it was necessary for all solid ingredients

[2] Resin obtained from the tree *Pistachia lentiscus.*

[3] Group of resins from Eastern Africa.

[4] Soft resin obtained from tropical trees from the *Burseraceae* family.

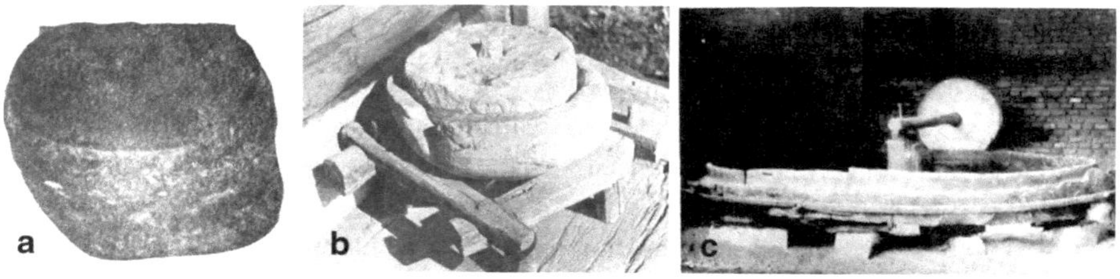

Fig. 8.1 Grinders used by artisans in the Romanian area: (**a**) "slab"—used in Transylvania to "rub" pigments to very fine granulation, (**b**) *agricon*—used to grind big amounts of pigments, (**c**) *oloiniţa*—drawn by horses and used to grind large amounts of solids. Photo: M. Mihalcu

(ochre, *ceruse* and *verdigris*) to "be ground dry and without leaving any out" [8]. The recipe for a "*poliment* learnt from the ancients" recommended to rub all the ingredients (chalk, iron trioxide, beeswax, and glue) "on a slab until they become sperm-like"[4].

For a red ink a Romanian manuscript [9] recommends for all the ingredients (cinnabar, a few drops of egg white, and gum arabic), after a 24 h "leavening", to receive the following treatment: "....rub them well until it (the ink) will flow well off the nib..." (Chap. 7). In a gold ink recipe [9] all the solid ingredients have to be "...ground on a marble (rudimentary mortar) until they become like water...". In other cases the recommendation for some or all the ingredients to be ground appears less explicitly but it is easy to understand.

The *hermeneias* indicate clearly sometimes that the components of a formulation have to be brought to the desired particle size separately. For an alcoholic varnish it was recommended [10] to grind each solid ingredient (*mastic, sandarac,* and *shellac*[5]) separately and then mix them. For another alcoholic varnish, made of very pure *sandarac, mastic, anime,* and camphor, it is clearly stated [11] "....and all these crush them separately each...". The same is indicated for a "Turkish varnish" made of *sandarac, mastic,* and *shellac* [3]. The grinding of the solid ingredients was done separately as well, on "the slab" (Fig. 8.1a), in the synthesis of cinnabar [6] and in the preparation of an alcoholic varnish having as solid components "very clean" *sandarac, mastic, elemi,* and camphor [12].

As we have already mentioned there are in the painter's manuals ink recipes asking to grind all the ingredients after mixing them [13]. Similar recipes ask for different processing of the ingredients: "..take all of them one by one and make them into a powder..." [3, 11]; "....crush them separately..." [3]; another ferro-gallic ink found in a manuscript [14] contained the same ingredients and recommended "...grinding them, sieve them separately...".

When successive "rubbings" were recommended, the target was two-fold: reaching an advanced fineness of the particles while achieving a thorough mixing of the ingredients. One such example can be found in the recipe for an adhesive for gilding, a *poliment*. In a first step, bole is ground on "the slab" together with soap,

[5] Alcohol-soluble resin, highest quality of *lac*.

with the obvious objective of grinding the former but also of thoroughly mixing the two components to facilitate the following technological step. During this step the mixture was mixed, in a pan, with melted beeswax and the resulting ternary mixture underwent a second "rubbing on the slab" during which egg white was added and worked well into the mix. The quaternary mixture was "rubbed" a third time on "the slab" until it displayed "butter-like" texture. At this point, little cakes were made of it which were stored and used as in Chap. 4. The same manuscript offers the recipe of an adhesive (*gulifarba* type) for which the solid ingredients (minium, *lime white*, ochre, litharge, and *verdigris)* were ground repeatedly (alternating with sieving). The fine material obtained was "rubbed" thereafter until it achieved the consistency of a very fine mortar (*migma*).

Two successive grindings were recommended in a cinnabar synthesis as well [8], one for the solid reactants (when litharge is used too, see Sect. 3.1) and one for the mixture obtained following the first heating (containing the black allotropic form of cinnabar). In both, grinding targets a thorough mixing and increasing the yield of the chemical synthesis. Two grinding steps were recommended when *lime white* was obtained from an old fresco and for the physical conditioning of pigments (one in the presence of water, the second in the presence of oil) [8].

For a "very good *poliment*" (see Sect. 4.2), after grinding the bole first and mixing it with melted beeswax, it was "rubbed on the slab", adding a "little" mercury and indigo [4]. The first step was meant to increase the adherence of the preparation while the second grinding was a guarantee for thorough mixing which would result in a uniform spreading of the adhesive preparation over the entire surface to be gilded. While the same manuscript contains recipes for adhesive preparations requiring three "rubbing" steps as well, it also contains a recipe for a *gulifarba* for which no grinding is mentioned, and several varnish recipes in which each ingredient was ground separately. One grinding step for all ingredients is recommended in [8] for an *ambole* and for two standard colors, *proplasma* (Sect. 3.4) and *sarca*[6].

8.2 Riddling, Sieving, and Straining

In the iconographers' workshops the separation of the granulometric fractions was done by riddling, sieving, or straining. Most of the time, pieces too big to go through were recycled and further ground. However, not too many details are offered in the *hermeneias* concerning how to conduct the grinding in relation to the requirements of the subsequent use of materials.

Riddling was done in a variety of riddles, from grates made of laths (used in lime slaking) to baskets made of twigs [8]. Big chunks of unhydrated material and gangue were separated from smaller ones when the mixture of lime and water was transferred from the vessel where hydration was conducted to a "slaker"

[6] Standard color used for the skin of characters appearing in compositions and portraits. Attributed to the legendary iconographer Manuel Panselinos and composed of ochre, ceruse, and a little cinnabar. Greek: *sarca*.

(a wooden vessel or a pit) where the hydration was completed during at least six months [3, 6]. In the case of lime, particles left unhydrated, when used in mortars for fresco, would be hydrated at a later time, increase in volume and result in "shots" (affecting the painted layer, Chap. 9).

When a finer granulometry was necessary in a raw material, screening was recommended. For some ferro-gallic inks this procedure was sometimes not required, at other times sifting through a flour sieve [6] or through a "fine sieve" [4] was recommended. Similar recommendations can be found for varnishes of different types [1, 15]. In the Romanian *hermeneias* there is no requirement to screen or sift pigments. Following their "rubbing" with the binders a reduction in particle size was occurring as well. It is easy to understand why a separation based on particle size was not possible thereafter.

Also within the topic of separations, mention has to be made of separation by filtration (straining) through a cloth for the cases when grinding was done in the presence of a liquid. Recommendations in the *hermeneias* are to use a "rag", a "fine sieve", or "a clean cloth" [16, 17]. Very seldom, as in the case of *ilinocopia*, an adhesive prepared from a single ingredient, the juice obtained by crushing garlic cloves (Sect. 4.4), there is no precise mention of how the filtration should be conducted ("strain it into a bottle" [16]).

The recommendation to grind with or without a liquid present was clearly stated. All the adhesives used in gilding were ground dry if the recipe did not include egg white. The same is the case for old fresco *lime white* (Sect. 3.3 and Chap. 9) [9], and for natural iron sulfate, a siccative [16].

The targeted granulation was indicated in the painters' handbooks more or less explicitly, from pieces weighing several kilograms, when processing lime and gypsum, to powders ("like dust") in the case of pigments. Adequate hardware was indicated and used. Whereas limestone and gypsum were crushed using sledge hammers, big hammers or horse-drawn grinders (Fig. 8.1c), cinnabar, ochre, and other pigments were ground in metallic mortars or, when subsequent steps required, were "rubbed" on a "slab" (Fig. 8.1a) or an *agricon* (Fig. 8.1b), a rudimentary mortar and pestle made of hard stone. How long the grinding was supposed to take was not indicated in the handbooks; it was part of the craft and it was taught by the master to the apprentice during the hands-on training.

8.3 The Media

There are cases (especially for some chemical syntheses) when, for the same chemical process, different ways to conduct the grinding of solid reactants are recommended in different manuscripts. For cinnabar synthesis, in a manuscript [8], after the first synthetic step (heating of sulfur and mercury and formation of a small amount of mercury sulfide) the resulting material was "rubbed on a slab with a little wine". For the same synthesis, another manuscript [18] recommended for both grinding steps (first that of the sulfur and mercury and then of the product formed) to be "dry". Dry grinding of the ingredients was used when the standard colors *sarca* and *proplasma* were prepared as well.

Maybe not all media used by the iconographers of Southeastern Europe are known today but, from what is known, it is clear that they were aware that any particular pigment cannot be ground in just any medium. In general, these media were surfactants which lower the surface tension, allowing the liquid to penetrate the solid material and complete the dispersion. Since materials were used shortly after grinding, conservation (spoilage) was not an issue.

Water was seldom recommended for grinding in the *hermeneias*. For the *lime white* obtained from old fresco mortar, after scraping off carefully the colored layer (with a knife or a glass shard), it was ground in a metal mortar in the presence of clean water. The particles obtained were air-dried and then "rubbed" on an *agricon*, in two steps. The first step was done in clean water, followed by air-drying and a second grinding without water. The second grinding was repeated, alternating with sieving, and returning to the *agricon* of the refuse left on the sieve (the same used in the peasant household to sift flour). Finally, the resulting pigment was sifted and "burnt" in an oven [4, 8, 12].

For colors in which linseed oil was used as binder and vehicle, pigment grinding was done in two steps as well. The first was done in the presence of water on the "slab" while the second, after air-drying, was done using "raw" (unsiccativated) linseed oil. Sometimes two successive "rubbings" in linseed oil were recommended [19]. Only *ceruse* was ground in walnut oil [3, 4, 8], as in the West European workshops [20, 21] and at Mount Athos [22].

The breaking up of gold leaf to obtain yellow-gold pigment was done by hand, using the fingers, in the presence of water, in a clean bowl. Sometimes a little gum arabic or honey was present too. Sedimentation was allowed and, subsequently, the supernatant was removed by decantation. The process was repeated three times. The purpose of the washing was to remove impurities and the gum arabic (when used) [3, 4, 18, 23, 24].

In [4], when grinding was recommended in a recipe for an adhesive, all the solid ingredients were included (gypsum, beeswax, ochre, bole) and then "...the rest is water and egg white." The same situation is found when cinnabar is used as pigment in red inks. It was "rubbed" with water [5] or with water in which a few drops of egg white had been added [6, 14, 16]. Both types of recipes contained gum arabic (adhesive) as well.

Grinding in the presence of water and wine was found in preparations of different materials: some standard colors (*sarca*, "the flesh") [25], an alcoholic varnish [23], and a variant for cinnabar synthesis [16, 18, 25].

In two red ink recipes from the *hermeneias* it was recommended for the cinnabar to be "rubbed on the slab" in the presence of wine, brandy and myrrh [6, 16] as in some West European painters' manuals [1]. Concentrated ethylic solution ("strong spirit") as medium for grinding is recommended in the recipe for an alcoholic varnish as well [16]. Grinding in alcohol was facilitating the process but was useful for the subsequent preservation of the varnish as well.

Another medium used in grinding was vinegar, used to "rub" the green pigment (*verdigris, grispan*) used in green inks (a copper acetate, Sect. 3.2) [7, 24]. Vinegar was also the medium used in crushing the fruit of the elder (*Sambucus*), used in the preparation of another ink [24].

Glue as medium in grinding was less used in Southeastern Europe than in the West. “Rubbing” in the presence of natural vegetal glues (from cherry or plum trees, Chap. 5), known and used in the West since Theophilus Presbyter’s time [26] and mentioned in Dionysius of Fourna’s *hermeneia* [22], was mentioned in the other handbooks only for some red ink recipes [4, 8, 14, 25].

“Limewater” (a saturated aqueous solution of calcium hydroxide) was used as medium to grind a white pigment obtained from dried calcium hydroxide paste (Sect. 3.3) and used in *fresco* [16].

Mercury was mentioned as medium in the grinding of an adhesive of the *ambole* type (Sect. 4.2) in which the solid ingredients in the recipe (bole, minium, tallow and ochre) were “…well crushed together with *quicksilver* ….and a little egg white.”

In rare cases grinding was done in the presence of two unusual media: human urine or snail slime [4]. Human urine was recommended for “rubbing” in only one manuscript [6], in the recipe of a blue ink. Snail slime (Sect. 4.5) was used in the grinding of solids for gold or silver inks in which it had the role of adhesive too [14]. In an Athonite *hermeneia* [22] alum is added to the same recipe.

A special conditioning procedure was used for rosin, obtained from the sap of conifers from the local flora, mostly fir tree (*Abies alba*). After the collected sap was melted and “boiled” (allowed to seethe until it became clear, without foaming), it was poured into a bucket containing cold water and pieces of different sizes were obtained. When needed, they were ground in metallic mortars or even on the “slab” [27].

Often, grinding in the workshops of Southeastern Europe was done without a medium which meant important savings of materials, work and time. In some cases it also afforded less pollution. Most of the time, when a medium was used in grinding, it was as an ingredient of the recipe, and not requiring additional separation steps.

References

1. Cennini CA (1960) The craftsman’s handbook, translated by Daniel Thompson. Dover Publ. Inc., New York
2. De Mayerne TT (2001) Lost secrets of Flemish painting including the first complete English translation of the De Mayerne Manuscript, B.M. Sloane 2052 (first ed.). Donald C. Fels (commentary and drawings), Alchemist, Hillsville
3. Dionysius of Fourna (1974) The painter’s manual. Translated by Paul Hetherington, St. Vladimir Seminary Press, Yonkers
4. Eastlake CL (2001) Methods and materials of painting of the great schools and masters. Dover Publications, New York
5. Grecu V (1936) Carti de pictura bisericeasca bizantina. Ed. Glasul Bucovinei,Cernauti
6. Mihalcu M (1984) Romanian medieval values. Ed. Sport-Turism, Bucharest
7. Mihalcu M, Leonida MD (1986) The 7th Conference “Romanian Bibliophilic Values”, section Restoration—Conservation, May 23–34. Tulcea

8. Mihalcu M, Leonida MD (2013) The hidden face of form and color, second ed. Ed. Eikon, Cluj-Napoca
9. Romanian manuscript 1164, Library of the Romanian Academy, Bucharest
10. Romanian manuscript 1184, Library of the Romanian Academy, Bucharest
11. Romanian manuscript 1188, Library of the Romanian Academy, Bucharest
12. Romanian manuscript 1212, Library of the Romanian Academy, Bucharest
13. Romanian manuscript 1555, State Archives, Bucharest
14. Romanian manuscript 1632, Library of the Romanian Academy, Bucharest
15. Romanian manuscript 1795, Library of the Romanian Academy, Bucharest
16. Romanian manuscript 1808, Library of the Romanian Academy, Bucharest
17. Romanian manuscript 2035, Library of the Romanian Academy, Bucharest
18. Romanian manuscript 2035, Library of the Romanian Academy, Bucharest
19. Romanian manuscript 2151, Library of the Romanian Academy, Bucharest
20. Romanian manuscript 4206, Library of the Romanian Academy, Bucharest
21. Romanian manuscript 4353, Library of the Romanian Academy, Bucharest
22. Romanian manuscript 4357, Library of the Romanian Academy, Bucharest
23. Romanian manuscript 5188, Library of the Romanian Academy, Bucharest
24. Romanian manuscript 5769, Library of the Romanian Academy, Bucharest
25. Romanian manuscript 5952, Library of the Romanian Academy, Bucharest
26. Theophilus Presbyter (1979) On divers arts. Hawthorne JG and Smith CS (Translators), Dover Publications, Inc., New York
27. Thompson DV (1956) The materials and techniques of medieval painting. Dover Publ., New York

Chapter 9
Iconographers and the *Fresco* Technique

The term *fresco* typically describes the *buon fresco* process in which painting is done upon a wet lime-plaster wall with pigments ground either in water or a water-based solution only. When the wall dries, the painting (the fresco) becomes an integral part of the surface, which makes it more durable than superimposed decorations. It also has a matte (dead flat) finish which allows it to be viewed from all angles without undue glare or reflections. Another type of *fresco* technique, *secco*, has the painted layer applied on a dry surface. It is basically a surface coating and therefore less permanent than true *fresco*, the mechanical resistance of the product being lower. The means to paint frescoes are and have been through history as many as there are schools of art.

In general the *fresco* technique is, as Michelangelo said, "the most difficult and daring painting technique, the ultimate test for a painter". Many painters, renowned for their paintings done in oil and tempera, cannot produce great frescoes. This technique requires from the painter self-discipline, courage, craftsmanship, and ability to make fast decisions [1]. Experience is also necessary, as is a good understanding of this technique since the colors used look differently when they are wet than when dry.

Most Orthodox churches in Europe have interior painting (Fig. 9.1) but there are in Southeastern Europe churches which were painted on the outside too, in *buon fresco*, few in the Balkans and none in the Russian area. During the nineteenth century researchers like Duchesne [2] and Bayet [3] wrote that it was impossible to find well conserved exterior medieval fresco. In the Romanian area there are still such works, 500 years old, which stood admirably the test of time (Fig. 9.2). Besides the church of the Voroneţ Monastery (Figs. 1.1 and 9.2c) which is also known as the "Sistine Chapel of the East", there are several others (Suceviţa, Humor, Moldoviţa, Arbore) located in the eastern part of Romania, which were painted in the fifteenth and sixteenth centuries. They all have interior frescoes (as most Orthodox churches do), but are also fully painted on the outside. These exterior murals were admirably conserved in spite of a climate which is harsh, with large variations in temperature from one season to another and abundant precipitation.

M. D. Leonida, *The Materials and Craft of Early Iconographers*,
SpringerBriefs in Materials, DOI 10.1007/978-3-319-04828-4_9, © The Author(s) 2014

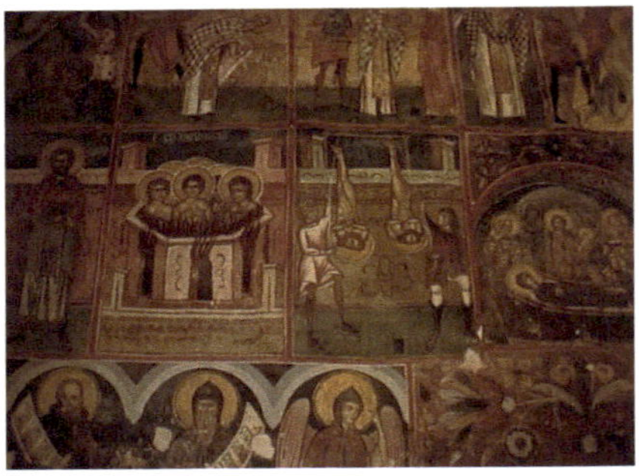

Fig. 9.1 Interior fresco (church of the Humor Monastery, in Northeastern Romania). Photograph: M.D. Leonida

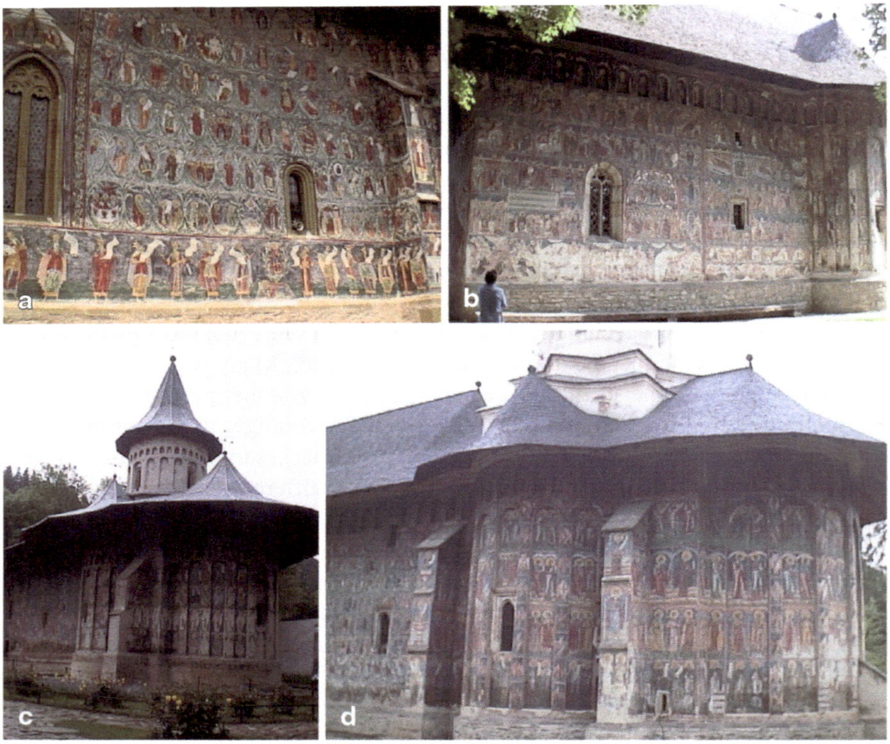

Fig. 9.2 Exterior frescoes in Northeastern Romania: (**a**) the South wall, church of the Suceviţa Monastery (painted in 1585). (**b**) the South wall, church of the Humor Monastery (painted in 1535). (**c**) the East wall, church of the Voroneţ Monastery (painted in 1547). (**d**) the South-East side, church of the Moldoviţa Monastery (painted in 1537). Photos: M.D. Leonida

Before actually painting a fresco on a wall (which could be stone, brick or a mixture thereof) some technical steps had to be completed:

- bricks with low or zero porosity were eliminated (based on the color of the vitrification spots appearing when calcination was conducted at too high temperatures) [4];
- inorganic salt deposited on bricks were removed;
- exposure of the wall to the air for longer periods of time (starting with the second half of the fifteenth century). Prior to that period deficiencies had been recorded, due to settling (walls "working" after being erected) even though wooden beams were often used as inner reinforcement.

Beginning with the first decade of the sixteenth century, in order to use a new church while the walls and foundation were still "working", temporary decorations (summary painting done using only 3–4 colors) were made. It was at least 5 years later that the permanent frescoes were painted. The prolonged exposure to the air seems to be beneficial if we note that the majority of the Romanian medieval frescoes do not have fissures in the mortar as do better-known frescoes (e.g. the superb frescoes in the famous Church of the Savior at Chora Monastery, in today's Istanbul, dating from 1315–1321, which needed extensive restoration work).

The mortar used in Romanian frescoes, as in Byzantine ones, was different from the one used in masonry. Natural calcium carbonate (sometimes even chalk was used) of a convenient purity has always been available and using wood only for burning (firing) contributed to maintaining that purity. That is why the *hermeneias* and other written documents from that time are very laconic regarding the purity of this raw material and its calcinations [5].

The white pigments used in frescoes were prepared using technologies which are still in practice and show a good knowledge of the resources, accurate observations done during generations of artisans, and a solid technical tradition which recorded and transmitted the information. We discussed in Chap. 3 (Sect. 3.3) these preparative technologies with emphasis on slaking of lime, and compared what the artisans did in Southeastern Europe with information from other areas.

Due to the high quality of the lime obtained, the fresco painters in this area did not resort, as did the Russian painters [1], to applying a first mortar layer in which a resinous binder was introduced or to consolidating the mortar layer with big nails as in some Italian frescoes of the fourteenth and fifteenth century [6].

The wetting of the walls prior to painting was done differently in different regions. In contrast to Cennini's recommendations or those of the Russian and Byzantine painters, the wetting of the wall prior to painting was done in the Romanian area abundantly, 5–6 times for brick walls and only 2–3 times for walls made of stones. Concerning the wetting of the support, we must keep in mind that, due to the large areas frescoes were typically painted on, the resistance of the mortar layers was diminished. This was mainly due to the fact that before applying the last layer of mortar, the previous one had partially dried. The solution to this, especially during the warm season, was to wet with water only the surface on which the mortar which will eventually hold the painted layer was applied. Churches are still painted using this method in present day Romania and a handbook for painters published not long ago in the United States still recommends wetting before the application of each layer of the fresco [7]. At the time the Romanian fresco painters were using this wetting step in the process of painting murals, Western Europe was starting to use what we call now the *mezzo fresco* technique[1].

There are no recommendations in the *hermeneias* concerning the rugosity of the mortar layer which could mean that the average granulation of the river sand used was probably satisfactory. It is however mentioned how the control of the granulation of sand was done, an element which is absent from older handbooks [6] and manuscripts dealing with this topic [8].

The Romanian painter's manuals do not make any reference to the amount of water introduced in the mortar or about the consistency of the mortar. It was prepared by apprentices under the supervision of an experienced master. The amount of water varied from one master to another and, over time, from one epoch to another. This is different from Cennini's Handbook [6] where recommendations can be found concerning the thickness of the "equalization mortar" layer and the amount of mortar which can be prepared for a one time use. In most cases the proportion of lime introduced in the mortar recommended by the *hermeneias* was similar to that indicated by Cennini and Vasari [9].

Sometimes the mortar layer contained sand, at other times a reinforcement material (hemp, flax, stems of other plants, straws, leaves [10]), the same as in Roman, Byzantine or Russian mural painting. Fillers (such as chalk, tile powder, brick, coal) were sometimes used too while there were cases when use of both filling agents and reinforcement materials was recommended. We have recently met some cases in which the intermediate layer of mortar, not the one on the brick or the exterior one, was very friable due to the excessive amount of sand.

When hemp, flax, and stem of other plants were used as filling agents, after removal of the wooden parts, they were cut to "standard" lengths and "conditioned" by soaking in a solution of calcium hydroxide. In some manuscripts, this treatment was replaced by the recommendation to use the mortar (to which the reinforcement agents had been added) only 3 days after preparation (which would have the same effect).

It seems that trends would come and go concerning the materials to be used in frescoes as in any other field. For instance, without any apparent reason, a reinforcement agent was being used very frequently (e.g. flax in the Moldavian frescoes of the fifteenth century) while in other regions of the country, at the same time, that same material was very seldom used. Numerous examples offer proof of the ability the artisans of those times had in choosing their materials in order to obtain a strong and durable mortar. An example thereof can be seen at a monastery in Argeş (Romania) where the exterior mortar layer was reinforced with leaves (low volumetric density) for the case when the lime used had mediocre binding characteristics.

[1] In the *mezzo fresco* technique the painting was not done on freshly laid mortar but on one that was partially dried. The result was that the colors penetrate it much less. In the appendix to the second volume of his "Prospettiva dei pittori e architetti", published in 1700, Ignazio Pozzo called *mezzo fresco* the true fresco technique (since it was the one in general use during that time in Western Europe). He wrote: "Be sure not to begin painting until the lime has reached a stage where it is hard to make a finger mark in it".

Irrespective of the type, the reinforcement material increased the drying time of the mortar the fresco was painted on. That is why, unlike their Italian contemporaries, the Southeastern *fresco* painters were able to apply the mortar layer on larger surfaces. The leveling was done as the work progressed. This reduced to an acceptable minimum the risk of visible differences between surfaces painted on different days. Another consequence was the increase in the time allowed for work from 10 to 14 h. In any case, the local climatic conditions contributed to the increase of that period of time too, compared to Italy for instance.

Regarding filling agents for the mortar layers, those used at Mount Athos (at the school of the legendary Panselinos, Chap. 1) were used over the entire area together with others with higher porosity and low volumetric density (like coal dust). This filler—the earliest used—was found in the remains of an old fresco (thirteenth–fourteenth centuries) in the Argeş district[2]. Chalk dust was also used as filling agent by the Romanian *fresco* painters. This filler was recommended by Vitruvius too [11]. Unlike in the ancient Roman frescoes, the Romanian artisans did not use marble dust in the final layer.

The total thickness of the applied mortar strata (8–20 mm) does not seem to be "standardized" and, in most of the cases, was lower than in the Byzantine frescoes. This was the case when the support wall was made of either brick or stone, unlike in Giotto's School. As everywhere else, in this geographical area the thickness of the mortar layers varied from one part of the building or wall to another. However the adherence between the mortar layers was excellent.

The number of mortar layers in the Romanian medieval frescoes (2–3) resembles the Byzantine (2) or Russian ones and it is inferior to the ancient ones (6–7 in Roman frescoes). Besides a variable proportion of lime in the different layers of mortar, the quality and quantity of the filling agents was also variable. In this respect, the Southeastern European frescoes are similar to the ancient ones and different from the Russian frescoes. As in the Byzantine murals, the appreciable dimensions of the frescoes led, at a point, to the necessity of simplifying the technique used in building the fresco support.

The chromatic values and vivid hues maintained over several centuries prove that the iconographers of times past knew how to choose their pigments. They had to resist exposure to light and to the atmospheric agents (when applied to the exterior walls of churches).

The painters' knowledge and ability in using pigments are evident not only when speaking of the resistance to microclimate and the alkalinity of the calcium hydroxide solution but also when considering the very method of application (how, how much, and to what extent each pigment changes its hue after the fresco dried). Concerning the resistance to the basicity of the environment a Romanian manuscript states that "…pigments have to stand the strength of lime, have to be long-lived and pleasant to look at…" [10].

Mihail Alpatov, whose opinion was that, in the context of a whole work, the colors are "the vehicle of spiritual expression", looking at the exterior frescoes of

[2] N. Constantinescu—personal communication, Archaeology Institute, Bucharest, 1980.

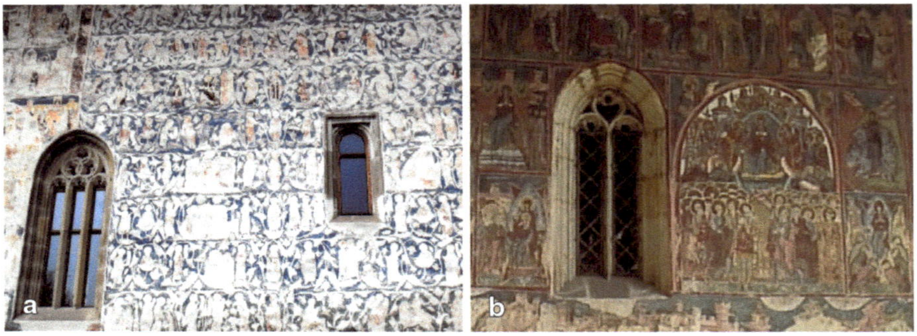

Fig. 9.3 Exterior frescoes at: (**a**) Probota Monastery (painted in 1532). (**b**) Humor Monastery (painted in 1535). Photo: M.D. Leonida

Northeastern Romania more than five decades ago said: "What an astonishing courage! The Romanian masters painted the exterior walls … without being afraid of the fact that the colors ….could fade" [12].

However, a question remains about them: how come all of them have been well preserved (Voroneţ, Humor, Putna, Moldoviţa, Suceviţa, Arbore), only those at Probota were far more poorly conserved? The climate is the same, since they are in the same part of the country, and Humor was painted only 3 years after Probota (1535 vs. 1532) and it is in a much better state (Fig. 9.3). A possible explanation for that (mentioned by a well-known art critic [13]) was that less durable technical procedures were used on the frescoes of the latter, the first church to be fully painted on the outside in this geographical area.

In some of the *hermeneias* a more or less complete list with pigments recommended for use in fresco was given and, in a few cases, even a list with those which should not be used was given. Sometimes it was indicated which pigments give certain hues when used in certain parts of the fresco or which pigments should be applied on a certain background (e.g. it was obligatory to apply a blue pigment on burned umber and on black). Some of the pigments used were natural (taken from the surrounding nature or bought), others were prepared in the workshops (Chap. 3).

The technological indications referring to the latter category of pigments contain an intermingling of precise details and, sometimes, recommendations that we cannot accept or understand today. Such an example can be found in a Romanian manuscript [14] where some components listed for the preparation of a blue pigment are not appropriate for use in *buon fresco*. At the end of the recipe, the anonymous author indicates the bibliographic source ("so do the Russian ones"). Immediately after that, in the same manuscript, there is another recipe for the preparation of the same pigment, a simpler and clearer recipe, adequate for the purpose and used in the painter's workshop.

In many cases, in the recommendations from the *hermeneias* the dosage of the reactants, the duration and the thermal regime of the preparative reactions and other details are clearly stated as is the method to control the preparations. Thus in the same manuscript, when presenting the preparation of synthetic cinnabar (using a mixture of mercury and sulfur), after everything is made clear, the final recommendation states that the heating should be made with oak charcoal only. In Macarie's *hermeneia* [15], a variant transcribed by a monk Eleazar (from Caldaruşani Monastery), when reference is made to the preparation of cinnabar by the same "dry" method, after indicating the necessary reagents and the quality control tests, it is stated that the heating should be made with charcoal obtained from a different wood essence. In both cases the working temperatures are higher than those absolutely necessary and the duration recommended for the reaction (24 h) seems to be excessive. Regarding the choice of the fuel and the strict specifications concerning its qualities, we have to be cautious today. For melting sulfur and properly heating mercury it is not necessary to use a certain type of wood.

Concerning the actual painting technique the recommendations available in the *hermeneias* display refinement and precision. Thus, when applying some colors containing iron oxides in their composition it is recommended first to lay on a pigment which mixes better with water. Unlike the Italian fresco painters of the fourteenth–fifteenth centuries, the Romanian ones did not modify the binder by introducing egg, casein or milk. The west European painters also chose to apply some costly pigments (like lapis lazuli, azurite) *a secco* (after the fresco had dried), held together by glue rather than egg yolk [16].

While some recommendations are similar, the painter's manuals of Southeastern Europe show differences from one region to another as well. For instance, indications found in the Romanian manuscripts are similar to those used by schools of painting from other countries with which the Romanian painters were in contact. An example is the preparation of the background recommended for painting figures (containing black pigment mixed with green, yellow and white) which seems identical with those recommended by Cennini [6] or Dionysius of Fourna [17] for the same task. At the same time, unlike Panselinos' school [18], with which the Romanian painters were in agreement through Dionysius of Fourna's writings, for some backgrounds the Romanian artisans used cinnabar mixed with lime white and ochre.

The preceding examples oblige us to draw other conclusions as well. One is that the *hermeneias* were compilations. Their authors—like their contemporary colleagues—had acquired data not only from local tradition and from the schools they had come into contact with but they recorded and kept using their own observations and adaptations, the result of a centuries-long activity and of a selection process competently performed.

The black pigments used in *fresco* were obtained by controlled partial combustion of numerous wood essences (oak, birch, tree bark), of vine shoots, bones, nut shells, or peach kernels (Sect. 3.4). Not all the details and justification for the uses assigned to each type of black have reached us through either written or oral testimonies. However some of them are known from different sources. In his manuscript Macarie of Căldăruşani [15] recommends to use a black obtained from the wood of a resinous tree to paint the pupils. He argues that "if you use the one you work on wood and canvas, the color will fade". Other *hermeneias* mention that a darker purple hue, like the one used to paint gray hair, is achieved by mixing black obtained from oak tree with a very well-specified white, obtained from the carbonated lime from old fresco mortar. At the same time a lighter shade of black used for more luminous backgrounds was prepared from vine shoots [19].

White pigments were prepared either from lime (slaked for a long time, or a shorter time, or from mortar of old frescoes separated from the painted layer and subsequently ground) or from chalk dust (Sect. 3.3). Although the compositions of the white pigments were very similar, like the black ones, the uses for each of them were different. It was known at the time that pigments with the same chemical composition could lead to different intensities as the sizes (and respectively the granulations) of the particles differ (Chap. 8). Each pigment transmits its own characteristics to the color it is used in. Thus, in painting the faces, it was recommended to use a white obtained from a calcium hydroxide slaked for a long time. The white obtained (by a technology described in minute detail) from the mortar of an old fresco was recommended for mixtures of pigments used in painting backgrounds, gray hair and human flesh. This white pigment was made through a process during which the filling agents (from the old fresco mortar) had not been removed. There are also instances when the recommendation is to use white pigment, without any other specification (e.g. in painting the folds of clothing).

Sometimes specifications were made regarding the use of the same "standardized" hues or colors for a certain part of the fresco. The white from the mortar of an old fresco was mixed with cinnabar or praseolite[3], while that obtained from recently slaked lime was mixed only with cinnabar in most cases. The chromatic accords and the great resistance over time also proved the pertinence of the centuries-old observations.

The green pigments typically used in fresco were obtained from hydrated silicate of aluminum, magnesium and iron (the popular name found in *hermeneias* is praseolite), malachite[4], or basic copper acetate (prepared from vinegar and metallic copper using a technology described in Sect. 3.2). The appearance of a basic copper chloride in small amounts in the green color is due either to its formation during the preparation of the copper acetate (due to the accidental presence of chloride in the calcium hydroxide solution) or to its later formation, following a double exchange reaction.

Yellow pigments were first obtained from a natural "earth", limonite[5]. Some of the pigments were from local sources, others were imported (e.g. the Constantinople ochre). Their covering power and their hue were very skillfully used (technically speaking). By calcinations—briefly described in the *hermeneias*—new hues were obtained from a different type of ochre. Unlike in Western Europe, in Southeastern Europe these new hues seldom had special names.

As red pigments, minium and cinnabar (natural or synthesized in the painter's workshop) were used, unlike the West European artisans who did not use the former

[3] Hydrated silicate of aluminum, magnesium and iron. Can be found as a variety of altered iolite of a green color and greasy luster.

[4] Basic copper carbonate. Fine, clear yellowish green, not reliably permanent when used as pigment. Found as major copper ore and, later, synthesized as Bremen blue. Used in jewelry too.

[5] Mixture of hydrated iron oxides. Found naturally as iron ore.

in *fresco* [20]. The synthesized cinnabar was not purified as was recommended in Theophrastus' book [21]. Also as red pigment, some aluminum silicates tinted by iron oxide were used.

It is interesting to note that the iconographers knew that cinnabar and minium were two different compounds and that they behaved differently when exposed to the atmospheric agents. Skillfully, knowing their pigments, the *fresco* painters recommended in a *hermeneia* (considered to be a compilation of Panselinos' teachings and Theophrastus' book) to use minium only when painting the red of the cheeks. At the same time, for lips, they recommended either a mixture of cinnabar with ochre and black or cinnabar with lime white, or cinnabar with burnt umber[6] and black [10].

The brown pigments were made of natural or burnt umber mixed with black (in a 1:1 ratio), and with some iron oxides either natural or roasted. They were used for the shadows around the eyes, nose and arms, etc [22].

The most frequently used blue pigments were natural azurite[7] and *lapis lazuli*[8]. Concerning the *lapis lazuli*, some *hermeneias* pointed out that it had to be mixed with some indigo and with calcium carbonate or lime white for better resistance over time. "Voroneţ blue"[9], has been added to the lexicon of art alongside colors such as the "Titian red" and "Veronese green". The magnificent fresco representing the Last Judgment, painted on a "Voroneţ blue" background and placed on the West wall (Fig. 1.1) of the church at Voroneţ Monastery is known world-wide.

In most cases the colors were prepared by controlled grinding the pigments with water or a solution of calcium hydroxide (Chap. 8). As in Cennini's book, it is often mentioned in the *hermeneias* that the more thorough the grinding is, the more adequate the color would be ("... grind it fine, to butter consistency...").

For a better "acceptance" of the color by the last mortar layer the *hermeneias* pointed out that if a thin skin of calcium carbonate had been formed on the mortar, openings were made in it with the trowel before applying the colors. In this way the water from the wet mortar seeped out becoming part of the pigment vehicle. Following the chemical reaction with the carbon dioxide from the air and formation of calcium carbonate, the painted layer becomes entrapped in its crystalline structure.

The appearance in the Romanian *hermeneias* of some vague indications regarding compositions of some color preparations to which egg or a boiling bran solution had been added led to the controversy (at the beginning of the last century) whether all the painting considered *buon fresco* in the Romanian mural painting was actually true *fresco*. Tempera, unlike the old Russian fresco where it was used for thin superposed layers, represents, when present in the work of the Romanian painters, retouching or later restoration.

---

[6] Reddish-brown pigment—it is obtained by calcining umber, a greenish-brown natural mix of manganese (major) and iron oxides.

[7] Basic copper carbonate. It is found as a minor ore of copper and it is used as pigment and ornamental stone, in jewelry.

[8] Sodium calcium aluminum silicate sulfate. For centuries it was the most expensive pigment. It was replaced by synthetic ultramarine.

[9] A color obtained by applying azurite on a black (charcoal) background (to enhance color intensity and insulate the pigment from the alkalinity of lime.

A certain waviness in the surface of the *fresco* mortar seems to be due to a technique by which the improvement of the general aspect of the work was attempted. The surface seems to be intentionally wavy, targeting a more vibrant light effect, achieving both a matte finish as well as the transparency of watercolor in some instances or oil or pastel reflections in some other instances. There is neither in the Romanian old painters' manuscripts nor in the memories of the painters still active any specification about the use of technology to obtain a fresco with a very polished, smooth and tough surface. We also did not find even vague recommendations like Cennini's or Battista Alberti's[10] regarding the finishing of the last mortar layer by means of a wet brush or the handle of the trowel.

The thin brushes (round, stiff or pointed) used in *fresco* painting by Southeast European artisans were made of mule mane hairs, mule jaw hair or hair above the hoof of the ox (the round brushes), and also from goat and squirrel straight hairs (the pointed brushes). Badger hairs or hairs from the tail of a cow did not appear to have been used in this area as was the case with Russian frescoes.

Combed hairs, very well cleaned and of the same length were tied with a thin thread and then inserted into the feather of an eagle or another bird. In case the hairs were not straight, the *hermeneias* suggested straightening them by means of "yellow sticky earth". The big brushes were made with pig hair. There are no recommendations as in Cennini's Handbook to use white hair only. Before being tied to a wooden handle, the hairs were either tied with a thread or glued with melted conifer resin. All brushes were kept during work in, and later cleaned with, a dilute calcium hydroxide solution.

Fresco is a form of art requiring a thorough knowledge of materials and high technological skills in order to make an art work which can stand time and weather agents (especially when painted on exterior walls). The passing of time and poor restorations have worsened the condition of many a mural in different countries. In countries with limited resources (as those in Southeastern Europe) restoration of religious monuments has been scarce and the elements have been the only enemy. These frescoes date from as early as the 1300s and are a testimony to the talent and skills of the artisans from that area, and to their success in producing masterpieces which passed successfully the test of time in places with a trying continental weather pattern.

References

1. Kiplik DI (1952) Pictura monumentala. Editura de stat pentru literatura si arta, Bucharest
2. Duchesne LMO (ed) (1886–1892) Le Liber pontificalis: texte, introduction et commentaire. E. Torin, Paris
3. Bayet CMAL (1908) Precis d'histoire de l'art. Picard, Paris

[10] Alberti was a theoretician: he gave a scientific basis to works of art and placed painting, sculpture and architecture on the same level as literature and philosophy. In his writings the craftsman became an intellectual.

4. Glen-Gery brickwork design guide. http://www.glengerybrick.com/about/manufacturing/firing.html
5. Mihalcu M (1984) Romanian medieval values. Ed. Sport-Turism, Bucharest
6. Cennini CA (1960) The craftsman's handbook (Translated by Daniel Thompson). Dover Publ. Inc., New York
7. Gottsegen MD (1993) The painter's handbook. Watson-Guptill Publ., New York
8. Pliny the Elder's chapter on chemical subjects (1929) Bailey KC (ed). Longmans, Green and Co., New York
9. Vasari G (1960) Vasari on technique (1960). (Translated by Louisa S. Maclehose). Dover Publ., New York
10. Romanian manuscript 2151, Library of the Romanian Academy, Bucharest
11. Vitruvius on Architecture (1931–1934) Granger, F. (editor and translator). G. P. Putnam's Sons, New York
12. Alpatov M (1982) Early Russian Icon Painting. Olympic Marketing Corp; First edition, Hopkins, MN
13. Comarnescu P (1961) Indreptar artistic al monumentelor din nordul Moldove. Regionala Creaţiei Populare, Suceava
14. Romanian manuscript 1795, Library of the Romanian Academy, Bucharest
15. Romanian manuscript 5769, Library of the Romanian Academy, Bucharest
16. Procacci U (1968) The great age of fresco. Giotto to Pontormo. The Metropolitan Museum of Art, Art Editions "Il Fiorino", Florence
17. Dionysius of Fourna (1974) The painter's manual (Translated by Paul Hetherington). St. Vladimir Seminary Press, Yonkers
18. http://www.macedonian-heritage.gr/HellenicMacedonia/en/B2.2.2.4.1.html
19. Leonida MD, Mihalcu M (2001) Black organic pigments used by medieval painters. The 34th Middle Atlantic Regional Meeting. Towson University, Towson
20. Zucchiatti A, Prati P, Bouquillon A, Giuntini L, Massi M, Migliori A, Cagnana A, Roascio S (2004) Characteristics of early medieval frescoes by -PIXE, SEM and Raman Spectroscopy. Nuclear Instruments and Methods in Physics Research B, 219–220: 20–25
21. Theophrastus (1774) History of Stones (*De lapidibus*). English version Sir John Hill, editor, C. Davis, London
22. Mihalcu M, Leonida MD (2013) The hidden face of form and color, 2nd edn. editor Eikon, Cluj-Napoca

Chapter 10
Icons on Glass: Materials and Technique

Peasant icons on glass offer a strange image of a folk art, clearly outlined and unique in its technical quality and sensitivity. While a few paintings on glass were found in Central Europe (mainly Bohemia and Austria), the area where this type of craft was practiced extensively and reached maturity was Transylvania and Northern Moldova. The beginnings of this artistic Romanian pursuit are not well known. Because of the frailty of the glass support, and of the poor adherence of the color to this shiny, non-porous surface, many, far too many, ancient icons on glass have disappeared over time. Icons found in different collections, which bear an indication that makes it possible to date them, belong usually to the second half of the eighteenth century (Fig. 10.1) [1]. In the Romanian area this genre reached its pinnacle between 1830–1900. After 1900 this technique was used by very few iconographers [2].

Some researchers consider that this form of art, painting sacred figures on the back of a glass plate, came to the Romanian area from Central Europe [3, 4]. Why did scholars and the higher ranks of the Romanian priesthood despise or even forbid their use? In agreement with the hypothesis of the above mentioned researchers, it is worth mentioning that in Transylvania most of the names of the colors, pigments and other materials used by the peasant painters have a Germanic resonance. Could this be rather an indication of those who supplied the materials? The themes and models used are designated by pure Romanian terms. According to the principles of iconography, the themes and their frequency as well as the characteristics of the details do not point to a Germanic model but to a Romanian, Byzantine or Near Eastern one. However further questions arise. Why did painting on glass penetrate into and why was it practiced and appreciated in the Romanian area whereas the cohabiting ethnicities of the same Catholic faith as the peoples of Central Europe did not adopt it? Why do the chromatics and golden background that give a personal touch to each color and suggest the serenity and magnificence of the beginning of the world, remind us of the Byzantine art? In Byzantium the art of painting on glass was known several centuries earlier [3], in Italy it was practiced since the fourteenth century, and sometime over the centuries it developed into a peasant craft too.

The icons were transported over shorter distances by carts belonging to the iconographers themselves, by those of peddlers specialized in the trading over fairly restricted and well delineated areas, or carried by shepherds during the seasonal

M. D. Leonida, *The Materials and Craft of Early Iconographers*, SpringerBriefs in Materials, DOI 10.1007/978-3-319-04828-4_10, © The Author(s) 2014

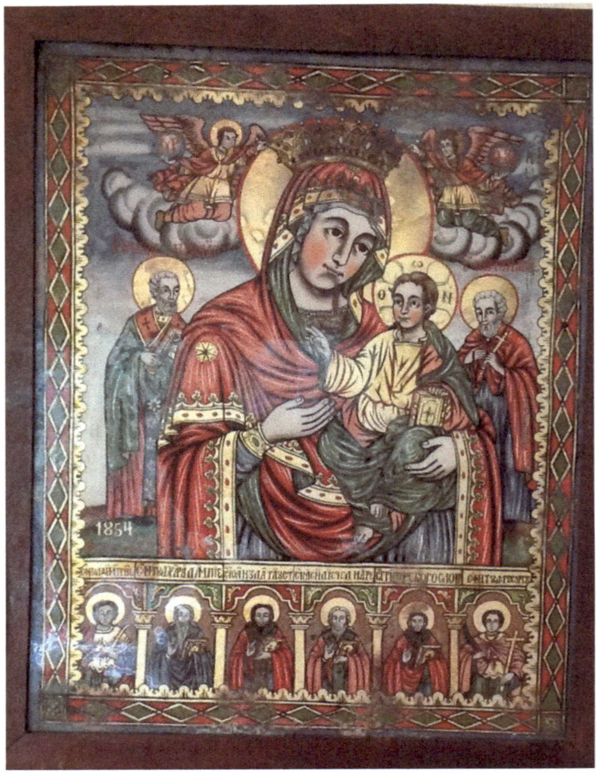

Fig. 10.1 Icon on glass, dated 1854 and signed. Photo: M.D. Leonida

migrations of the flocks from one pasture to another. Icons painted on glass spread throughout all the Romanian provinces, often far away from the centers where they were made. In the late 1970s we found in a house located in the southeastern region bordering the Black Sea two such icons painted in the remote Transylvania (at Nicula, well-known center for icons on glass). Icons were bought as presents or for a girl's dowry because, in certain regions inhabited by Romanians, during certain periods, no girl could marry without owning an icon. Icons representing the Virgin (Figs. 10.1 and 3.1) or the Nativity were especially popular since they were considered symbols of fecundity.

The most numerous, some well-known, centers where iconography on glass developed were in Transylvania, whereas those in other regions were few and much less known and studied. The products of the various centers bear certain local characteristics that can be easily detected [5].

It may be assumed that in almost all the centers, the icon painters on glass started to paint based on the skill and knowledge acquired in their previous activity using other techniques, or learned from those mastering this technique during that late medieval period in the region we study. Certain findings have shown that a number of church painters also painted icons on glass [6]. The imprint of professional painters on wood and the iconographic indications of the *hermeneias* can be clearly detected in the icons painted on glass. In many of the centers producing icons on glass, such as those in Southern Transylvania (*The Olt Country*) for instance, two different currents were detected: one represented by the skilled professional iconographers and another one represented by much less educated craftsmen [7]. This is not surprising since long before the art of painting on glass developed in the Romanian area, some remarkably skilled iconographers had lived and worked there.

The waning of this art happened as discreetly as its appearance when the conditions, thoughts and feelings of the communities to which it was addressed changed, in the first decades of the twentieth century. Today, this form of art, wrought over centuries, can only be viewed with the fondness and devout admiration of descendants. Nowadays, modern knowledge, based on multiple artistic experiments, can discern the technique of this exceptionally interesting folk art and may study it without the prejudice imposed by the canons of academic painting. As for other types of icons, while iconographic aspects of the icons on glass were studied by a good number of researchers, little has been written about the skill and technical ability of the craftsmen who painted them.

The following pages represent an attempt to fill in this gap based on information from the painter's manuals together with those collected from some old iconographers who painted on glass and were still active in the second part of the last century. We shall discuss the raw materials and the materials used, certain preparatory technologies and, finally, the techniques used by the Romanian iconographers who used glass as the support of the pictorial layer.

In an iconographer's workshop the helping hands were generally represented by the family, but within its members there was a certain division of the labor like the one occurring in the other techniques of painting as practiced in that area during the same period of time [8]. The steps that required no artistic gift or any special technical skills (grinding and sifting of the pigments, their "rubbing" in a mortar with a pestle, the preparation of adhesives, dye stuffs and additives, and of binding materials, and making the frames) were carried out by members of the family with lesser experience and qualifications. The coloring and the application of the local hues were carried out by more skilled members of the family. Still more qualified ones would make the drawing, the writing, and put in the patches of light. In almost all cases, the retouch—almost impossible in this technique—and the application of the gold leaf were done only by the head of the workshop. In workshops with a high production the colors were applied concomitantly on several icons, both to spare materials and to use time efficiently.

The peasant iconographers worked mostly during the cold season, when work in the fields and forests and shepherding slowed down. The icons were sold at fairs or by peddling, for money or exchanged for cereals or other wares from the homestead of the buyer. In many instances this trade took place also during winter when peasants and shepherds stayed home and many weddings were celebrated [6].

The materials used in Romanian icons on glass were: glass for the support, for the picture itself—pigments, dyes and colors, adhesive materials, binders, additives, gold leaf, and several wood species for the frames and the backing of the icon.

The glass was of the soda-lime (calco-sodic) type and it protected the painted layer in the same way varnish does on easel painting. Glass was obtained from small factories (*glăjării*) existing, as a rule, in zones covered with forests due to the large volume of wood needed for the furnaces. Glass was also supplied by local merchants, by hawkers or by pilgrims coming from different places to monasteries on the patron saint's day [6]. The glass was cut into small pieces of various sizes for easier transportation.

In smaller glass factories the technological process for glass making was a rudimentary, age-old one. A mixture of raw materials[1] was melted and then blown into cylinder-like shapes, the thickness of the wall being approximately the same as that of the future glass sheets. This thickness was not the same throughout the cylinder or throughout the glass sheet obtained when the cylinder was split lengthwise along a generating line and "stamped" down on slabs of smooth chamotte bricks, previously heated in special ovens. As neither the melted nor the "stamped" glass could be maintained at the required optimal temperatures, gas bubbles of various sizes remained within the glass (carbon dioxide resulting from the thermal decomposition of the carbonates in the raw material).

Since it was not possible to maintain the same temperature throughout the entire mass, inner stress lines appeared on cooling, increasing the brittleness to a great extent. Subsequently, long after the icon was completed and framed, the glass broke without any apparent cause, probably due to accidental mechanical stress (vibrations, mechanical shock, uneven dilation, etc). When these accidents occurred in connection with certain significant family events, they acquired a miraculous connotation.

In addition, due to the imperfect, rudimentary processing of the glass, the surface remained slightly undulated, having been cooled too much and too fast before being "stamped" down. On the older icons one can readily detect these undulations with the tip of one's finger.

The effect of these imperfections was not always unpleasant. The differences in the thickness of the glass, the undulations and air bubbles embedded in the material and the fact that the colors are seen through the glass impart, owing to complex optical phenomena, depth to the hues and result in vibrant colors. This imparts to the icon that unique, unequalled charm we have already mentioned. Hence, part of the magic of the old Romanian icons painted on glass may not be due only to the artistic genius and technical ability of the iconographer. When, towards the end of the nineteenth century, glass from commercial sources was exclusively used, without any of the defects mentioned, and when the Romanian folk painting on glass showed the first signs of decline, the charm of this kind of folk art was not the same anymore.

The size of the glass sheets used for icons was not standardized and various sizes were used by different artisans in different centers or even in the same center during different periods of time. In centers like Nicula (in the midst of a poor, less developed area) the icons were smaller than in the affluent center Scheii Braşovului,

[1] Raw materials used for manufacturing glass: the main ingredient quartz sand (silica), soda ash, limestone, and, sometimes, cullet (for furnace efficiency).

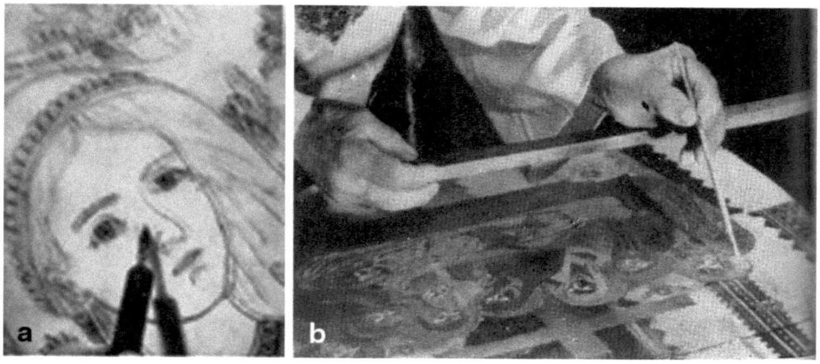

Fig. 10.2 Painting an icon on glass is a stepwise process. First step is the drawing: (**a**) an age-old pattern is applied under the glass and the painter traces the contour of the images on the future icon; (**b**) in the next step the artisan fills with color the shapes drawn in the previous step. Photo: M. Mihalcu

since the price of glass weighed heavily in the cost of the completed icon. No icons were painted on mirror glass in this geographical area.

Before painting, fat was carefully removed from the glass with lye wash. The glass was then rinsed with clean water and dried before painting. Lye wash was prepared either in large vessels or in the hollowed trunk of a hard wood tree by scalding ash. Many of the causes of exfoliation of the painted layers on some collection pieces are due to imperfect, uneven cleaning of the glass before painting.

In painting on glass the drawing was usually copied after a model (Fig. 10.2) or was drawn freely, without a model. The drawing was done in pencil, ink, China ink, black or red tempera, copying xylograph or chromatographic engravings and prints, engravings or pictures in the religious books available to the iconographers. Some artisans drew from imagination or from the details of several models or engravings so as to create a composition of their own.

When copying a model, the artist's contribution started with the elimination or introduction of some details, figures or ornaments done in such a way as not to perturb the composition. The skill, feeling and knowledge of the ancient iconographers could clearly be discerned, so that the use of models lost its standard character, the iconographer trying to make each icon into a unique piece. Some of the peasant icons, by ignoring the perspective, resulted in a charming evidence of a fascinating primitivism (Fig. 10.3). The unobtrusiveness with which these gifted art creators subjected the personal elements they introduced to the main theme is remarkable. It can be seen that, eventually, this led to drawings far removed from the original model, to a great variety of icons with the same subject, painted by the same artist, even when they were mass-produced.

Note has to be made, however, that some very gifted iconographers existed who seldom used models. By getting rid of the restriction of a model they were able, when inspired, to paint genuinely charming new compositions, sweet in their naivety. The models were made, were borrowed, were copied, were restored, refreshed, and inherited. Some iconographers possessed hundreds, others only dozens. They

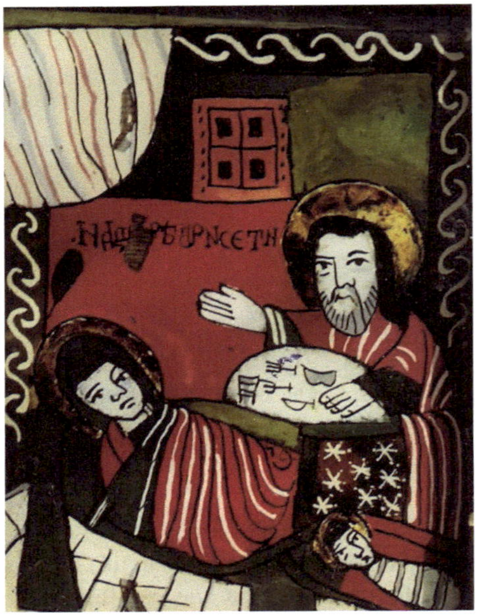

Fig. 10.3 Painting on glass from Transylvania (second half eighteenth century) in which the laws of perspective are ignored, giving the icon a very primitive look. Photo: M.D. Leonida

were lost by wear or accident (fire, flood, invasion, vandalism, etc.) so that few still exist nowadays.

The pigments used in painting icons on glass were obtained from the environment (sedimentary deposits in valleys, colored rock inclusions), from other iconographers' workshops, prepared in the artisan's workshop, or bought from merchants and peddlers.

Raw materials from natural sources (iron salts, manganese and copper hydroxides and oxides, cinnabar, kaolin, chalk) were processed (cleaned, ground, sieved and washed) in the peasant workshops by some of the less qualified members of the family. From the *hermeneias* and Romanian *miscellanea* it can be seen that, in the Romanian area, several preparative technologies used in other parts of the world were known and practiced with local innovations [8]. By using these technologies a number of excellent pigments were obtained from the lime white or white lead up to the synthetic cinnabar and green copper acetate (Chap. 3).

According to the formulas and personal preferences—difficult to reconstitute today—certain organic dyes obtained from plants (by infusion or extraction) were used in addition to the inorganic pigments, as the age-old tradition recommended. Today, however, only their use in dyeing textiles is remembered. These dyes were fated, by their very composition, to a more or less rapid decomposition under the sunlight which was not stopped by either the glass or the superposition of colors. These, however, offered some protection from the attack of the environmental agents to which icons were exposed. At times this attack was strong enough for the dyes to disappear completely. The chalky, earthy or even faded colors in some older icons, when not caused by the use of certain pigments of questionable quality, represent what is left from parts colored only using dyes whereas others, painted only using inorganic pigments, maintained the fresh brilliance of day one.

The black pigment used in painting on glass by the Romanian icon painters was prepared from carbon black obtained by incomplete combustion of certain indigenous resins and also by thoroughly grinding the charcoal obtained following incomplete combustion of certain organic materials (bones, oak or linden wood, grapevine stems, peach stones, walnut shells). With the pigment obtained from grapevine stems a cold grey color of particular beauty was obtained, characteristic for painters of yore from this area (Sect. 3.4).

Unequally used, the few white pigments had a fairly varied composition (fine chalk or kaolin powder, *ceruse* and—starting from the second half of the nineteenth century—zinc oxide). The white pigments obtained from kaolin or chalk were prepared from the raw material found in natural deposits by repeated breaking down, washing, drying and sifting (Chap. 8). White lead (*ceruse*) was bought, prepared in the workshop or obtained from other painters who prepared it by letting vapors of acetic acid (from concentrated vinegar prepared in the household) act on lead ore for about 40 days in the presence of carbon dioxide produced by the anaerobic fermentation of horse dung. When used in oil colors this pigment was "rubbed" with walnut oil in contrast to other pigments which were "rubbed" with linseed oil [8]. While these pigments were seldom bought, zinc oxide was always purchased from traders.

The red pigments were cinnabar (*vermilion*), minium, ferric trioxide, and, starting in the nineteenth century, fuchsia red. These were all provided by traders with the exception of cinnabar. This last pigment was obtained from natural local deposits and was also produced in some of the iconographers' workshops following recipes found in the Romanian painter's manuals (Sect. 3.1). This procedure was the same as the one used in medieval Europe, at Mount Athos, or in China two thousand years earlier [9]. Contamination with minium (of higher density and cheaper than cinnabar), a fraudulent trick, was far less harmful in painting on glass than in easel or mural painting (owing to the protection of the glass and overlay of other colors and of gold leaf).

Ferric oxide (iron trioxide), a dark red pigment, was obtained by calcifying ochre or directly from natural deposits. The tint differed according to the ore it was extracted from, its purity, and the temperature and duration of calcination. The higher the temperature and the longer the duration the darker the red, due to the higher proportion of magnetite (Fe_3O_4). As in easel and mural painting, iconographers replaced sometimes cinnabar by a pigment obtained from a mixture of colored ores (hematite and goethite) or a mixture of cinnabar and hematite (the latter due to its lower price) [8].

Yellow pigments were obtained from natural deposits or bought; however a yellow, as yet unidentified, plant extract was also used [10]. The ochre tint varied depending on the proportion of aluminum silicates in the ore, which would sometimes vary even within the same deposit. Ochres were particularly interesting for painting on glass because of their high opacity and covering capability. The imported *Thassos ochre* had a redder tint owing to its higher content of iron trioxide (Chap. 2).

The green pigment used was copper acetate which was prepared, as in Western Europe or at Mount Athos, in closed containers where acetic acid vapors (from concentrated vinegar) reacted with pieces of copper (Sect. 3.2).

Brown and purple pigments were manganese compounds obtained from natural deposits or bought from commercial sources. Azurite ($Cu_3(CO_3)_2(OH)_2$, Chap. 9) and ultramarine[2], the former from some natural sources and both from commercial sources, were used as blue pigments. Later on they were replaced by cobalt blue[3]and ceruleum[4], both commercially available.

Gold pigment, primarily used from the middle of the 18th to the end of the nineteenth century for aureoles, stars and sometimes for the background, was prepared from imported very fine bronze powder or gold leaf and was available commercially or from other workshops [8, 10]. The use of gold is unanimously considered to be the influence of Byzantine painting. The bronze powder was increasingly used after oil colors started to be used. In some icon centers (Braşov for instance) it was little used, in some others (Nicula) only seldom, but in Laz-Sebeş for instance it was used and at times overused. In the older icons and the simpler ones, and even in the entire work of some iconographers, no golden bronze powder or gold leaf were used but ochre was used instead (less expensive).

Starting with the middle of the nineteenth century, in some icon centers (*The Olt Country*, Laz-Sebeş) the background of the icons on glass was no longer painted. Instead, metalized paper or very thin tin foil was applied on the back of the painting. Since during the manipulation this background foil slipped and the colors did not adhere very well, the painted layer soon began to peel off. The outlines however resisted better since the colors used to draw them were prepared with a stronger binder. During this period the number of colors used to paint on glass was reduced to a selection of three to five colors (white, light red, green, ochre, and gold [5]).

At first icons on glass were painted with tempera colors and the influence of painting on wood panels was obvious. In the Nicula center, at the beginning, watercolor was also used but far less often [10]. Tempera colors were prepared by "rubbing" the pigments with a pestle in rudimentary mortars made of hard stone, with an emulsion of animal glue (from sheep skin or rabbit bones), egg yolk, and small amounts of ox gall and vinegar. Vinegar was added to the binders as a coagulant and preservative of the colors against biodegradation. The iconographer Ilie Poenaru-II remembered [5] that in the second half of the nineteenth century the colors were "rubbed" with a solution of bone glue mixed with very little linseed oil.

Analysis of the painted layer of icons painted in tempera colors in the early days of some of the icon centers (e.g. Nicula) is almost impossible since most of them were subsequently restored using oil colors. Later on, at times that varied from one center to another, oil paints started to be used. They were prepared by "rubbing" the pigments with linseed oil. *Ceruse* was the only one which was "rubbed" with

---

[2] Ultramarine—a complex mineral, a sulfur-containing sodium-aluminum silicate of variable composition ($Na_{8\text{-}10}Al_6Si_6O_{24}S_{2\text{-}4}$).

[3] A blue to green pigment consisting of a mixture of cobalt oxide and alumina, in variable ratios.

[4] A greenish blue pigment containing mainly cobalt stannate.

walnut oil because it is itself a pretty strong siccative (Chap. 8). With the other colors a colorless siccative (usually copper acetate) was added after "rubbing".

The technique of painting on glass was very simple [6], like that of traditional potters, or Easter egg painters, and it was the same throughout the entire Romanian area. However, in icon painting colors had a more conventional purpose than in other techniques, for they were supposed to be the vehicle for an idea. The parts delineated by the drawing were generally covered by the same bright hues, underlining the decorative and graphical character of the genre.

Four main difficulties were encountered in painting on glass. First, because the painting was done in a reversed position, as in a looking glass, the side painted on became the back of the icon. Hence, it can be easily understood that the order of the colored layers was reversed, thus making it fundamentally different from any other technique. The painter started with the drawing, then continued by writing the names, painting the luminous aspects and ruddy spots on the faces, introducing the local hues, the background, and finally the gold aureoles, the stars and gold leaf background.

A second difficulty was the virtual impossibility to retouch. Thirdly, colors have a poor adherence to glass, and the last difficulty is the fact that icons on glass cannot be restored when necessary as easily as those painted on wood or those painted in the *secco* technique on dry plaster.

The brushes used by the iconographers to paint on glass were of various sizes, depending on necessity (drawing, writing, contour of the ornaments, highlights, application of the local color, and background). Writing and tracing of the contours was done with brushes made of cat hair, sharp cherry wood slivers, goose quills, and later on with metal nibs. At other times, the brushes were made with squirrel, polecat, or badger hair, or with hairs taken from above the mule's or the ox's hoof [8] or even with young children's hair, taken from above their ears [3]. Straight or straightened hairs were used, tied together with waxed thread, cut to equal length, and introduced into a goose quill which had previously been left to swell in water. On drying it tightened and held the hairs. When a wood sliver was used, the tuft of hair was fixed in place also with waxed thread wound around it at one end. In some instances, for better adhesion, the tuft of hair was first introduced in melted fir tree resin.

To start painting, the glass was put down on a flat surface. If a model was used, it was inserted under the glass (Fig. 10.2), face downwards—the back of the model towards the glass and, to see the drawing better, it was first oiled with fat or kerosene to become transparent. In rare cases the drawing was inverted, the model not being oiled but put directly under the glass, face upwards. If the model had faded, the outline was enhanced with the so-called "glue ink" (sheep skin or bone glue solution mixed with dry carbon black).

At first the outlines were always traced with "glue ink". Later on, the same black pigment was introduced in a mixture of linseed oil, turpentine and a little lead acetate (a colorless siccative to ensure quick drying). In a rare occurrence, a peasant woman iconographer from Nicula, was known to have used instead of carbon black finely powdered linden or fir tree wood, incompletely burned [3]. The drawing was also done, though less often, with a tempera type of color prepared with a dilute emulsion of egg yolk in which carbon black was suspended. In certain centers (Nicula and Banat), some of the painters drew the outlines in black and other lines in red, especially the contour and features of the face, hands or uncovered parts of the body, as in *fresco* painting.

After drawing came the writing—the names of the saints, the title of the composition and, less often, the names of figures or objects difficult to recognize. In many instances the painter could not write and the letters were drawn in a decorative motif. As such it was interestingly placed into the composition and the chromatic harmony of the painting. After many and repeated copies of a model, the letters and the words copied by artisans who could not read themselves had no longer any meaning.

The abbreviations that often appear on Romanian icons on glass were imposed by the restricted space, for the painter's convenience, or because the composition could be easily identified due to the main figure. The abbreviations became more numerous after the Latin alphabet began to be used in the Romanian Principalities[5].

Sample pieces of glass were used for testing the colors, the mixtures thereof, and the colors' adherences to the support.

The characteristics of painting on glass in general as well as the primitive technique of the iconographers made it difficult to create shadows. In many cases the shading of a color and the attempt to give an impression of volume was realized by doubling the contours with a white line or with a shade darker than the outlined color, or with white dots and small squares. There were also centers where no attempt was made to enhance the value of the color. Some painters graded the colors and did not use the pure unmixed pigment except with certain local colors (red, green, and ultramarine). Why that is the case remains an open question. The color was shaded on the uncovered parts of the body (in grey), on grey or yellow garments, and the façade of buildings.

Other painters (from the Lancrăm center) who graded the green, the light brown and the delicate light grey of the faces were less successful in their endeavor. However, these few attempts at grading color as well as the strange perspective and architectural details attest to the influence of the technique of painting on wood panels.

On some icons shading appears to be due to the deliberate or accidental application of layers of color of different thickness, to the thinning of layers by atmospheric agents, or to the repeated, unpropitious dusting and wiping with a cloth moistened in kerosene or in the oil burnt in oil lamps. It should be mentioned however that, due to their marked ornamental aspect and their votive use, grading was less important in icons painted on glass, which makes an extended discussion unnecessary.

In almost all cases when oil-based colors were used, the painting was covered with a protective layer of turpentine [11], of linseed oil or mixed linseed oil and turpentine. This was a kind of varnish, but applied on the back of the painted layer. Its role was to better seal in the colors and to protect the pictorial layer against humidity

[5] Before 1860 on the territory of today's Romania the Cyrillic alphabet was used to write in the Romanian language. It was officially replaced between 1860–1862 by the Latin alphabet.

[3, 5, 11]. This protective layer was sometimes applied even when tempera colors had been used.

The frames of the Romanian icons painted on glass were made of more or less durable wood, with or without a profile, painted, colored or stained with tanner's drench. The wood most used was fir but maple, plum tree, oak and walnut were used as well. The frames were sometimes made by the artisan himself or by the village carpenter. The icon was transported after having been framed as the painting readily came off. The back of the frame was covered with thin wood boards that both protected the icon and rendered the frame more rigid. In the Nicula center cherry tree bark was also used to cover the back of smaller icons [3]. The papers put between the painting and the back for increased protection often offered interesting indications regarding the age of the icon, or about certain techniques used, or were even of historical or social interest.

Until the end of the nineteenth century the frame was assembled by wooden pegs and later on by metal nails. In some cases the framework was joined together by wooden tenons, which increased its rigidity and stability.

Only in the eighteenth century Lancrăm, eight frames made by "Nicolae the Painter" had the back piece and the frame cut as one piece from a block of wood, the glass being slipped into place along a groove cut between the actual frame and the back piece [3]. The size of the frames was, obviously, adjusted to that of the glass. Frames were stained or decorated with oil or tempera colors which also protected them. As a rule, grey, black, dark blue, brown or purple were the colors chosen. At any given time in any certain center, preference was given to a specific color. Thus, for instance, in the Nicula center frames were painted blue with red stripes up to the end of the nineteenth century. Around the beginning of the twentieth century the great majority of frames were painted in black and later on they were only stained. In the Iernăuţeni center, however, all the icons which have come down to us have the frames stained.

It is no longer possible to detect how the icons on glass were initially decorated and protected because they were repeatedly cleaned, restored and repainted. Some icons present flowers, geometrical designs or even certain figures painted on their frames in a questionable taste and skill, in strong contrast to the general color of the frame. These decorations on the frames must have been painted more or less in the same way as those on peasant furniture. In some centers (Şcheii Braşovului, for instance) the frames were decorated by tracing brown, wavy lines with a "comb", in the same way as on the Romanian pottery [12].

Although this genre, the painting on glass, was imported from other geographical areas, the Romanian iconographers developed it according to their own artistic tradition and brought it to heights recognized by researchers in this field from all over the world. Their originality and unity of style are impressive and the instinct shown by the choice of colors tells of accomplished artisans making their mark in the Southeastern folk art of times past.

References

1. Wendt HC (1953) Rumänische Ikonenmalerei. Eine kunstgeschichtliche Darstellung. Erich Röth Verlag, Eisenach
2. Ioanidu IC, Radulescu GG (1942) Icons on glass. Bulletin of the Commission for Historical Monuments, fasc. 113–114
3. Dancu I, Dancu D (1975) Peasant painting on glass. Editura de Stat pentru Literatura si Arta, Bucharest
4. Iorga N (1934) Les arts mineurs en Roumanie, vol. I. Édition de L'Imprimerie de L'État. Bucharest
5. Mihalcu M, Leonida MD (2009) Din tainele iconarilor romani de altadata. Ed. Eikon, Cluj-Napoca
6. Irimie C, Focsa M (1969) Romanian icons painted on glass. Ed. Meridiane, Bucharest
7. Mureşanu A (1977) Reflexii (Restituiri). Ed. Dacia, Cluj-Napoca
8. Mihalcu M (1984) Romanian medieval values. Ed. Sport-Turism, Bucharest
9. Mihalcu M, Leonida MD (1981) Romanian medieval techniques for the synthesis of HgS. Bul Fiz Chim 5:242–245
10. Popescu AI (1969) Arta icoanelor pe sticla de la Nicula. Ed. Meridiane, Bucharest
11. Irimie C (1971) Icons on glass. Ed. Meridiane, Bucharest
12. Mihalcu M, Leonida MD (2013) The hidden face of form and color, 2nd edn. Ed. Eikon, Cluj-Napoca, p. 244

Chapter 11
The Apprenticeship of an Iconographer

Much has been written about icons, with a great deal of interest and remarkable competence, especially during the twentieth century. However, almost all that has been written has been about the visible face of icons. Those who studied them were, in most cases, art historians and art critics, regular historians or theologians interested in Biblical history and in the part of the icon conveying the message (implicitly in the iconographic programs). Naturally, they went as far as the limits of their respective specialized fields, and no further.

Much less—almost nothing at all—was researched and published about one of the unseen faces of icons, that of its functionality and about the way it can be achieved. Nonexistent also are studies concerning the relationship between Church-icon-iconographer, a relationship which is intimately connected to the above-mentioned functionality.

This has always been true, not only during times when, in the name of a vulgarizing materialism, in some areas of Southeastern Europe authorities were hostile to the propagation and practice of any religion and, implicitly, to religious art, but always during history. Little was written due to other reasons as well. The scarcity of the sources of information is one reason. Another, as we already emphasized, is the lack of interest theologian researchers showed, at least in their written contributions, for this subject. Although references in written testimonies and oral tradition are not numerous, they exist and they could and should be utilized.

Before looking at the training of iconographers of Southeastern Europe, some questions have to be asked and answered. How did the Church and the iconographers perceive the icon? What was its functionality? In what conditions, or how and by whom could an icon with a certain functionality be made? In which way and conditions was the activity of the icon makers to take place, in order to achieve a "working" connection with the Transcendental? Which was the modality to obtain and preserve the sacred gift (*charis, har, Grace),* a different one from the one needed to paint, a quality considered absolutely necessary for a creator of sacred art? And finally an answer is required to the question: what was the relationship between the Church and the iconographer? Was it only of the type client-provider (the one who orders something custom-made and the one who provides it) in the most complex sense?

M. D. Leonida, *The Materials and Craft of Early Iconographers*,
SpringerBriefs in Materials, DOI 10.1007/978-3-319-04828-4_11, © The Author(s) 2014

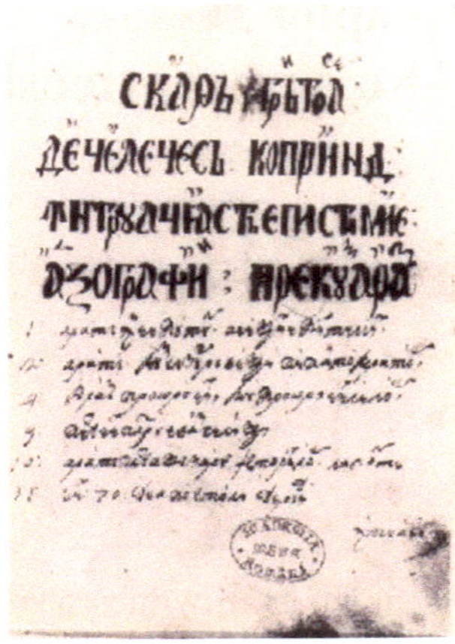

Fig. 11.1 Table of contents of a Romanian *hermeneia*. Photo: M. D. Leonida

The icon—the product of the activity of iconographers—was and still is more than a prop needed for practicing a certain religion. Because it is a means to communicate, to form a complex bond, it needs to be efficient. However, this efficiency depends on intimately knowing a series of factors involved in the process of making an icon. Once they are known, more and more interesting details become apparent and lead to a better and more detailed picture and understanding of the material and spiritual life in a certain geographical area during a certain period of time. These factors, together with the knowledge about materials and about the processes to obtain them, and with the craft involved in making the icon are parts of what we may call the hidden face of an icon.

11.1 The Apprenticeship

The written sources of information concerning the apprenticeship of iconographers are the *hermeneias* (Fig. 11.1), especially their prefaces (foreword, *predoslovia*) [1]. In these manuscripts we also found sporadic information, appearing as annotations scattered among other notes (Fig. 11.2). Sometimes it was concerning certain family events. Within the sources of information presently available, the *hermeneias* appear as unique. Along the line of technical tradition we can say that no new information can be obtained any longer. The religious painters connected to some extent with the old iconographers' tradition have disappeared since long. Also, during the last two centuries some practices—not only spiritual ones but technical ones as well—from the religious painting have been neglected more or less and, ultimately, eliminated and forgotten through the succession of generations of clergymen and iconographers.

In connection with this process note has to be made that, although some points of view, regulations, and attitudes have origins during the Apostles' and Church's Founding Fathers' times, they seem rarely observed today throughout the Orthodox East. Probing the subject of the old rigors and of the deep faith of the medieval

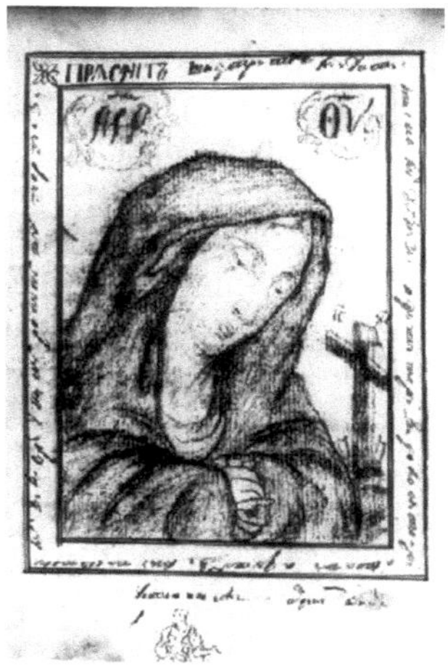

Fig. 11.2 Annotations and drawing done by a temporary owner (possibly the painter Gh. Tattarescu (Romanian painter. Started as an apprentice of a church painter (his uncle). Later he obtained a scholarship and studied at Accademia de San Luca, in Rome. Gh. Tattarescu was the co-founder of the National School of Fine Arts, in Bucharest (1864). He taught there for many years and painted many religious frescoes and icons, besides paintings with non-religious subjects.)), on a page of a Romanian *hermeneia*. Photo: M. D. Leonida

painters was less than welcome, even disturbing when an iconographer with more than 30 years' activity was interviewed [2]. Without providing details, the person told a story about the time when he was painting a church. He had to fire one of the subcontractors because "he was praying" while on the scaffolding and thus was providing "a bad example for the others". The painter did not seem aware of the danger that, following this action, the painting might lose one of its most important and significant components.

Numerous and varied reasons led to this situation in the Southeastern European religious painting. Among others, the mutations appearing in the spiritual and material life during the last three centuries have to be mentioned. Among them the effects of social utopias, new "faiths", in the name of which devastating "crusades" were conducted, as well as the industrialization with its consequences are not the least important. In the Eastern European area, the domination imposed on an important part of the world during the second half of the past century by a social system based on a vulgarizing materialism had a stronger impact.

However, this process started all over the Orthodox East long before, and the social phenomena just mentioned were not its starting point. At Mount Athos for example, the very heart of Orthodoxy, there were visible signs thereof since the beginning of the nineteenth century, as mentioned in the preface of the first Athonite *hermeneia* by the researchers who discovered it [3].

Among the few existing written documents concerning our subject of interest in the Romanian area during the twentieth century, worth mentioning are the memoirs presented during a public conference [4] and published in a newspaper article [5] by the iconographer Costin Petrescu. In them he reminisces about his childhood and adolescence spent in the studio of his father, an iconographer himself, descendant of a long line of such painters. It is clear from these memoirs that, since the last quarter of the nineteenth century, the consecration to which the Romanian icon painters were subjected (the process through which the *Grace* was bestowed on them, enabling them to create a functional icon), together with the continuous psychological self-conditioning (to maintain the *Grace)* were already no longer adhering anymore to the old controlled, and rigorous ritual. In old times they had been mandatory for the formation of an icon painter. Memories about this process persisted in this geographical area, within the circle of iconographers, even during the second part of the nineteenth century.

All over the Orthodox East, while the auto-conditioning and, even more, the consecrations continued to be practiced one way or another, their charismatic content decreased continually, their objectives gradually changed, and their lay component increased. Gradually they were becoming something else.

The consecration, for instance, was becoming more of an acknowledgment of a training process, the official recognition of the right to practice the craft of painter/church decorator and painter of icons with sensibly diminished functionality. This was because a few prayers, said by no matter whom, did not have the same weight or the effect in bestowing the *Grace*, as they once had. The event became, like many others, a festivity at the beginning and the end of the training for whichever technical or artistic occupation.

There is another reason too for the fact that there are so few written materials concerning the subject of interest herein. Following an age-old way of thinking, information concerning the activity of iconographers was considered secret and could only be transmitted within the circle of icon painters, directly from master to apprentice either orally or through practical demonstration. The affiliation with that particular circle was obtained, as we are going to see soon, through a consecration ceremony, during which the *Grace* was bestowed.

The information concerning how to obtain this sacred gift needed by each iconographer, together with that concerning iconography and iconographic programs, were considered secrets. The same was true for these artisans' craft (knowledge about raw and processed materials, preparative technologies and painting techniques). In this respect, Costin Petrescu, noting the relationship between these "secrets" and the Transcendental world, specified that "the Romanian iconographers from times past … considered the iconographic matters as well as the technical ones real pastoral secrets" [5]. He also mentioned that the master was "obliged to transmit (to the apprentice) all the knowledge which had been transmitted to him, without hiding anything, with the belief that God will reward him and that here, on the Earth, the parents will thank him and honor him as befits his honor." Obviously this obligation concerned not only the knowledge received by the master, traditionally, from his own master but also the knowledge acquired through personal effort, during his entire activity.

When the future iconographer was accepted as apprentice, this had the significance of an initiation, and an admission into the closed circle of those privy to the secrets of the icon makers. As we already mentioned, this way to look at things is very old. It has been around since Antiquity, through the early Middle Ages in the western world as well as in the area which lived, for centuries, under the Byzantine influence. The foreword *(predoslovia)* of a Byzantine *hermeneia*, translated from Greek and published by the Byzantinologist Vasile Grecu was addressed—like the eighteenth century one, authored by Dionysius of Fourna—"to all the students in painting and only to them".

This approach lasted in the workshops of the Romanian iconographers longer than in other areas of the Orthodox East; it was abandoned a few decades after it ended in the Russian area and several centuries after Western Europe [6]. From the *predoslovias* of the Romanian *hermeneias* which were preserved for us, it is clear that this way of transferring the knowledge an iconographer needed was still practiced around the middle of the nineteenth century in Southeastern Europe, and it was restricted to the closed circle of the painters and their apprentices. Moreover, after this transfer started to be done in writing, it was still restricted to the same circle of initiated. Thus, in the foreword of a *hermeneia* [7] the author was addressing very specifically "to the brothers and to those who start in this craft", that is, to painters, and in another *hermeneia* [8] to "the knowledge loving apprentices". A third manuscript [9] was addressing the content to those "loving and eager for the weariness from painting" and, finally, a fourth one [10] was stating that those with whom the "secrets" are shared are "son and brother painters".

To see how closed was their circle and how they were regarding it, it is relevant to look at the rapport existing among iconographers and between a master and his apprentices, respectively. In the foreword of a Romanian *hermeneia,* the author clarifies the rapport existing between him, the painter, and his apprentices by addressing them as "most beloved and pastoral sons" [7]. Addressing his peers in the same manuscript the anonymous author calls them "brothers and those within this divine craft…". It is easy to notice that, in both statements, the author considers his painting, as well as that of his colleagues as having a sacred character and the relationship between master and apprentice as having another dimension than those spanning other technico-artistical crafts. To further clarify these ideas, in the same preface, the author, taking another position was recommending "….and do not do your task poorly and no matter how but fearing God, and with good devotion, as a godly thing". By mentioning censorship imposed by a supernatural power the idea of its involvement in the entire process of creating an icon becomes clear.

It comes across from the painter's manuals that iconographers considered, and rightly so, that the icon belongs to sacred art and that it has more dimensions and a much more complex functionality than the products of activities in secular art. Accepting this makes it easier to understand and justify the need to have consecration ceremonies for apprentices and journeymen, as well as other obligations iconographers had and to which they totally subjected themselves for the purpose of continuous mental self-preparation (auto-conditioning), during their entire activity. This was true for any icon painter, whatever his rank, abbot, apprentice, journeyman. The latter mentioned activity, that in connection with maintaining the icon maker in a state of *Grace*, typically intensified before a consecration ceremony as well as before starting an important piece (e.g. painting or restoring the frescoes of a church, painting a series of icons, etc.).

According to the tradition of the Orthodox Church, the first icon miraculously appeared during Jesus' lifetime[1]. According to the same tradition, the first icons

[1] Orthodox tradition records that Abgar Ukhama, the ruler of a small kingdom (Edessa, the first Christianized kingdom in the world, in the second century, was located between the rivers Tigris and Euphrates) received, brought at his request, by his servant Anania, Christ's face miraculously imprinted on a kerchief, He had used to wipe his face. With it and through it Abgar recovered par-

made by hand were made by Luke the Evangelist[2]. If we take this into account, together with the fact that the documents of the 7th Ecumenical Synod state that the first icons appeared during the Apostles' time, we realize that the authority of icons was traced back to the authority of the Apostles.

The Church considered that the painter is the one who, using a language specific to art in general and to painting in particular, conveys a message to the soul of the person who looks at and worships the icon. He tries thereby to make a connection between the world of mortals ("the seen one") and that of the Transcendental ("the unseen one") without the image having anything idolatrous in it. The function of the icon remains as a relay. It is a real testimony of the iconographer's faith, it realizes its potential and functionality as propagator and efficient go-between (relay) only when, through the message it sends, it does not conflict with the dogma.

Otherwise said, the painter, through the icon he makes, has to not only describe the Heavenly Kingdom but also to connect him who prays in front of it. It was considered that the icon ("the visible of the Invisible" as St. John of Damascus called it) was a means for the believer who worshiped it to break away as much as possible from the earthly and to fully contemplate what is pure and above the human condition. To produce such a piece of sacred art, the iconographer needed to possess additional qualities which are neither necessary nor compulsory for the lay painter. Only by having them (obtained by consecration, maintained by continuous auto-conditioning) can the iconographer impart the quality of relay—to use a modern technical term—to the sacred art object created by him.

Costin Petrescu, the iconographer we already mentioned, recalling the echoes of the old tradition of the Romanian Orthodox Church and of the painters who adorned it with that extraordinary *Biblia pauperum* and visual theology, says that these artists considered themselves "...the keepers of the Holy Gift inherited from the Apostle Luke to whom traditionally the painting of the Immaculate Virgin is attributed" and also that they considered "the information and the technical knowledge as true pastoral secrets"[5]. The foreword of a *hermeneia* [10] clarifies not only what the gift of an iconographer is but also its source ("...Godly gift from far above given" and subsequently "...God given and ordered"). In the *predoslovia* of another painter's manual the same matter is further clarified in a prayer used during the consecration service, which we already mentioned. In this prayer the Evangelist Luke is clearly invoked [8].

We emphasize here that, in the outlook of religious painters of times past, the "Holy Gift" we already mentioned was a gift, it could not be acquired otherwise but through a consecration (ceremony through which *Grace* was bestowed). This could

tially from leprosy, the terrible scourge of Antiquity. This kerchief (*Mandylion*), an icon result of a miracle, is first mentioned in a sixth century document, and indirectly by Eusebius the bishop of Cesarea in his writings too. Thereby, it is considered by the Orthodox Church that Christ Himself has endorsed the making of His image.

[2] References to these early icons (all of Mary Mother of God) can be found in liturgical texts as well. In there it is also mentioned the fact that Mary bestowed on them *Grace* and miraculous powers. This was another way to implicitly show that the *Grace* and the specific powers of the Icon are transmitted to all the faithful copies of these archetypal icons.

only be accomplished by a man of the cloth, a priest, that is, an individual who not only had the *Grace* but also had the power to transfer it. Following this consecration, the painter did not subsequently have either the capacity or the possibility to transfer, in turn, the *Grace* to a third party. As a consequence of achieving the state of *Grace*, the painter was becoming a potential creator of the functional icon (able to connect to the Transcendental world).

Similar ideas and statements appear in the Athonite *hermeneia* of Dionysius of Fourna. There was everywhere in the Orthodox East a common position concerning the way to look at and understand the iconographer, his activity, and the icon, as a sacred art object made by him. The European West, without totally overlapping with the conclusion of the Orthodox East, has a pretty similar outlook. Michelangelo, for instance, said that "good painting comes close to God, joins Him, and is only a copy of His perfection, a shadow of His brush". In this formulation too, the role of the religious painter as a creator of sacred art with a well-determined functionality is emphasized.

We do not have yet a clear and certain indication that the future iconographer, before starting his apprenticeship, was subjected to some tests in any way, concerning his natural inclinations, psychological profile, artistic aptitude, and obedience to the Christian regulations. It is very likely that a preliminary assessment of his family's respectability and of his own behavior in society was done. There is no reason to believe that the procedure was different in the case of future apprentice painters than it was for entering as an apprentice in other fields. For the time being we will go only as far as to state that at the beginning the gate to apprenticeship was wide open. Throughout the apprenticeship, as during the subsequent activity, thorough and conclusive tests were administered in several areas: quality and utility of the sacred art objects produced, social behavior, observance of the craft's values and regulations, and continuous auto-conditioning. This situation seems to be similar over the entire Orthodox East.

The situation in the Christian West was roughly similar. The exigencies were less numerous than in the East though. Cennino Cennini, when talking about the beginning of the apprenticeship of the future painter, says that "he who very much wants to master this art needs first to adorn himself with love, fear, obedience, and perseverance" [11]. All these points were, surely, verified by somebody. Note has to be made that the state of *Grace* is not mentioned either in the previous quote or in other West European testimonies. Practically, what Cennini mentions are general attributes required to practice any artistic or technical craft. Exception has to be made for fear which, probably, means "fear of God". As a conclusion, we can say that in the European Orthodox East admission to apprenticeship of a future religious painter required more qualities than in the European West.

The apprentice often was the son of another painter. Consequently he had had some contacts with the technical part of painting and with the psychological rigors to which the painters were required to subject themselves. Almost never would a painter have taken his own son as an apprentice "because, as a parent, he would not have any authority" as Costin Petrescu says. The future apprentice was given as "servant for room and board" to another painter. On this occasion "the parent was recommending to his colleague to provide not only teaching and food but beating as well when mistakes would warrant it" [5]. The master was required to give to the apprentice adequate nourishment for the physical effort asked of him and for his age, nothing surprising or different from what was required when apprenticeship in any other craft was starting. In the same spirit, we should note the severity of the treatment during the process. It was allowed to go as far as physical punishment when the apprentice either made a mistake or was disobeying his master. Concerning the latter, it is interesting to note that the forewords of the Romanian *hermeneias* mention nothing about it.

At Mount Athos [12] and in the rest of Southeastern Europe the beginning of the apprenticeship of a future iconographer, according to "a teaching to the one who will learn the craft of painting", was restricted mainly to copying repeatedly and for a long time numerous patterns (*isvoade*) made by some "accomplished teachers" or "skillful teachers". These were well-established painters, fully recognized as creators of sacred art. This first stage was mandatory. Naturally, during the apprenticeship the future painter was used for other, less demanding, tasks like bringing pigments and additives to a required granulation ("grinding the colors"), different weighing and mixing procedures, hydration of calcium oxide ("slaking of lime"), "black" mortar (used for warping and leveling) preparation, preparation of the reinforced mortar (with hemp and flax fibers, ground straw, other fillers), and conditioning of some materials. Other chores were not directly connected to painting (e.g. cleaning the studio or the workplace, transporting materials to build scaffolding, etc.). Only copying established models was a formal part of the first part of the apprenticeship and it is recommended as such in the forewords of the *hermeneias*.

The scope of the copying was to help the apprentice improve his drawing ability, to make him familiar with some requirements of religious painting, but also with the way to present and interpret portraits of saints and biblical scenes, in order to avoid slipping into heresy. This is why, in this stage of the apprenticeship, models made by previous masters recognized as correct ones, archetypes, were used. In the fourth century Evagrius Ponticus[3], in his "practical treatise" (The *Prakticos*) was saying very clearly that it is necessary "to endeavor to draw icons by always looking towards the archetype". During this stage of the training process, the master had the possibility to assess the artistic gift, the technical ability, and the general behavior of the apprentice.

Dionysius of Fourna's *hermeneia* mentions in connection with the beginning of the apprenticeship in the monastic republic at Mount Athos: "...make new trying to find some of the peerless Panselinos, to make your long apprenticeship over them" [12]. The reference to a well-known iconographer and to assiduous practice in drawing appears clearly here.

The practice for the beginning of the apprenticeship period in the Russian religious painting workshops was similar and went even further. There are several

[3] Known also as Evagrius the Solitary (345–399 AD) was a Christian monk and ascetic. He was a classical scholar, believed to be among the first to begin recording and studying the teachings of the Desert Fathers.

known cases in the history of art in this geographical area in which a young painter used in his activity the models of an established iconographer who preceded him. Rublev, for instance, in some of his big works made at the beginning of his career used some models of Theophanes the Greek [2]. In some old Russian breviaries[4] and some psalm books, miniatures were found that used as models decorations from older manuscripts and even patterns created by the same Theophanes the Greek.

While some types of "collaborations", identically motivated, have been recorded in the activity of the Romanian religious painters too [13, 14], they were less numerous.

In spite of similarities concerning the beginning of training of an iconographer, some regional differences can be noticed. In the Romanian area, differently from Mount Athos, the practice of using old patterns was starting before the apprentice had learned how to knowingly use the standard proportions of the human body (the "measures of the man"), the composition of Biblical scenes and the characteristics of saints' portraits. These "secrets", contained in the painter's manuals, were accessible to the apprentice only after a certain time following the first consecration. In the foreword of a Romanian *hermeneia* [15] this was clearly shown in the statement " (the apprentice)…to endeavor himself on the patterns, without concern for measurements…" and further, in the same manuscript, "…first to teach himself and to improve mindlessly, without measurements, in order to master the craft…". In this way it is indirectly suggested also that, during this stage of his training, the apprentice was monitored by his master in the area of drawing too. Even more, it was specified that it was necessary to check whether practicing on old patterns "agreed with him".

The standard proportions we already mentioned were considered of great importance and that is why, at Athos as well as in the rest of the Southeastern European area, the *hermeneias* were starting the section with technical recommendations by revealing these "secrets". They had the same content in all *hermeneias* in which they appeared, even if sometimes they appeared under different names. Sometimes it is about "man's measures" [9] or about "how many measures the man's whole body is made of" [16]. In other cases it is simply said "the natural in human measures" [10] or "…the knowledge to mark the natural man's body…" and "…the man's measures…" [17]. Finally, in other texts, the announcement that the discussion of these canons was following was made by saying "showing for the normal and natural measures" [18] or "the measures of the normal and natural one" [15].

About testing and monitoring the evolution of the iconographer—as a creator of sacred art in the Romanian *hermeneias* which have reached us, there are not too many clear details. Some were, probably, considered "secrets", others being too trivial and similar to those in other crafts to be mentioned. They still existed and there are some sporadic notes to this effect. It appears that even when, after the consecration, the apprentice was allowed among journeymen, he was still subjected to some tests required by his guild or patron.

A document from the eighteenth century, containing regulations concerning the activity of iconographers [19], established that, before a painter was commissioned

[4] Liturgical book containing canonical prayers.

to paint a church, the church authority was required to examine the way the painter worked ("…to his craft if it is skillful…"). Only after this examination and depending on its results would he get the commission ("was allowed"). Surely, on this occasion, the manner in which the painter was interpreting the Biblical texts (observing the dogma) was under scrutiny, not only his craft and artistic gift.

Similar situations were observed for other areas too. For instance, the Ecclesiastic Assembly, held in Moscow in 1551, imposed upon the Russian iconographers similar monitoring and testing. In the documents of this Assembly it was said that "God's icon shouldn't be left in hands which deface or disgrace it". In this quote there is a clear reference to both artistic skill and obedience to the dogma. Additionally, it was specified that an icon should not be painted by "hand unadorned with Godly gift" [20]. The reference to *Grace* appears clearly here too. It puts the icon painter on a higher level, with more requirements, than the lay painter.

11.2 The Consecration Ceremony

After the apprentice had gone through the preparatory phase of the training when he had his first contact with drawing, the passing into the next phase was marked by a consecration ceremony. Although in the sources of information available we did not find a reference to a preliminary period of fasting, prayer and meditation, followed by confession and communion, it is very likely that these stages were included. Taking into account the way these types of events were considered by the Church of the Orthodox East, it is not likely for them not to have been stipulated.

After having gone through a consecration and having benefited from its effects, the iconographers' apprentices had access to the secrets of religious painting, including the canons concerning the human body. The main sources of information about those secrets were teaching (delivered either orally or by practical demonstration by the masters) and the *hermeneias*. The latter, organized in the two parts (iconographic recommendations and technical ones), contained a good part of the theoretical and practical knowledge required from a religious painter [1].

The first consecration ceremony of the apprentice iconographer, performed only after the first instruction and testing stage had ended, was called the "prayer" in the *hermeneias*. We are not aware of the use of the term consecration during the centuries when this ceremony was considered necessary. In the following we shall use the term consecration because, practically, during this ceremony a transfer of *Grace* was effected. The person upon whom this was bestowed was not able to transfer it, in turn, to somebody else. Costin Petrescu, previously mentioned herein, uses "consecration" when he mentions this ceremony.

The fact that those who compiled the *hermeneias* thought along the same lines can be seen in how they referred to it in the forewords. In one of the *hermeneias* [15] it is clearly stated that the religious painters were "set into the Godly craft with *molifta*[5] and with *otpust*[6]", subsequently restricted from straying even in their imagination from the provisions of the dogma and the teachings of the Church.

The entire ceremony of consecration was taking place in front of an icon of the Virgin (*Hodegetria*) [8, 9]. This icon was asked to act as an intermediary in the relationship with the Divinity, the provider of *Grace*. This specification suggests not only the role of go-between but also that of adviser (the "Guiding Virgin"). Reasoning like this, we reach the definition of the iconographer as creator of sacred art and, implicitly, the relay function of the icon.

In the forewords of the Romanian *hermeneias* there is frequently mentioned a connection between Jesus Christ and the iconographer mediated by Virgin Mary and Luke the Evangelist, the patron of the Eastern and Western European iconographers. The same idea is mentioned in both the foreword of the Athonite *hermeneia* of Dionysius of Fourna and in Cennino Cennini's "The Craftsman's Handbook" [11], that is in both the Byzantine and West European areas. In Cennini's "Handbook" included on this ticket are St. Eustace—his patron saint—and "all the male and female saints from Heaven". Similarities like these evoke distant common sources of inspiration for the religious painting in both the Orthodox East and the Western area.

Requirements for the apprentice at this stage concerned the craft, skill, correct life (sober and according to religious and social regulations). At the same time the future apprentice needed a recommendation (materialized by escorting him to the consecration) from his master or the person who supervised the apprenticeship. This person was required to verify, prior to the ceremony, both the ability (artistic and technical) and also the behavior in the community of the apprentice.

In the forewords of some of the Romanian *hermeneias* there are details about the course and the content of the consecration ceremony of the future painter. The priest started the service by blessing the apprentice, and the prayers "Our Father" (invocation of the Divinity) and "Heavenly Lord" (invocation of the descent of the Holy Ghost unto all the participants in the ceremony) were said. At the same time, through the second prayer, the "enlightenment of the mind" was solicited with the purpose of knowing the Truth, achieving a pure mental condition, "straightening" the iconographer's entire future life towards Good and eliminating all slips into psychical impurity. The bestowing of *Grace* was also solicited from the third hypostasis of Divinity. All these were part of what today we would call "entering the bliss", without which the participants in the ceremony were considered still unprepared to achieve a "holy work". Following these prayers a few hymns (*tropare*[7]) were sung [9].

The first of them was a dogmatic imprecation ("may the mouths of the pagans who do not worship Your Icon become mute"), evolving from the way the Church

[5] *Molifta*—Prayer read for the atonement of sins in some special services in the Romanian Orthodox Church.

[6] *Otpust*—Prayer for liberation (conclusion of a ceremony) said at the end of an important Orthodox religious service.

[7] Religious hymns sung in church on different saints' feasts, during the year

was regarding the first icon realized by human hand (Evangelist Luke's). The imprecation addressed to the iconoclasts[8] showed, probably, that the respective *tropar* was created and introduced not long after the iconoclastic wave. It constituted and still constitutes a means to fight against those who contested the qualities and the functionality of icons.

The second hymn was one praising an event (Transfiguration) and, at the same time, one of penance and humility ("...may the purifying light shine for us sinners..." [8]). At the same time, in it, the descent of the Heavenly Grace unto the participants in the ceremony (especially the priest and the apprentices) was again solicited. At this point it is necessary to emphasize that the reference to the Transfiguration was emphasizing, once again, the consecration character of the ceremony, as a second sign in this direction after the prayer "Heavenly Lord" previously mentioned. It was at the same time a repeated solicitation for the descent of *Grace*. From this hymn the idea of the whole ceremony meant to create and ensure a direct connection for the future iconographer with the Transcendental appears clearly. The consequence of establishing this connection was that the apprentice was obtaining the quality considered to be the most important for a creator of sacred art.

In the third *tropar* (also of penance and humility) appeal was made to the Virgin to act as intermediary and to humbly ask from the Divinity "forgiveness of sins"[8]. Continuing, after mentioning the all-embracing love shown by Jesus when he endured the Passion in order to save the entire mankind, the hymn becomes one of praise (of the Savior).

The fourth *tropar*, after praising Virgin Mary, defines what the icon is considered to be by both the painter and the Church, thereby redefining and highlighting the purpose of the iconographer's activity. This aspect was previously discussed.

During the moments following in the ceremony, the priest was signing again above the head of the kneeling apprentice, asking at the same time the participation of the entire audience, and soliciting the descent of *Grace* ("Let us pray to God!"). The cross sign and the touch of the apprentice's head by the priest's hand (the person who was transmitting the *Grace*) marked the bestowing of the state of *Grace*.

The next step in the ceremony was an ample prayer specific for a consecration. Mentioned in Dionysius of Fourna's *hermeneia* [12] as well as in the Romanian handbooks, this is another argument in favor of an age-old common model preceding both the practice at Mount Athos and the one involving other areas of Southeastern Europe. This prayer, after an invocation of Jesus, was mentioning the icon miraculously imprinted on the kerchief (previously mentioned) and the one of Mary painted by Evangelist Luke. It also contained a solicitation for the *Grace* to be bestowed upon the apprentice ("...enlighten, fill with wisdom the soul, mind and heart of Your servant [the name followed] and guide his hands to draw innocently and with good pleasure the impression of Your Face and of that of Your Immaculate Mother....").

[8] Iconoclasm—deliberate destruction of religious icons and other religious symbols, usually for religious or political reasons. People who engage in iconoclasm are called "iconoclasts". More generally, it came to define any individual who challenges established dogma or conventions.

It is easy to see in this excerpt the role as a connection relay creator the iconographer had during his entire activity. We note here the concise definition of the psychological state the painter needed to be in during the entire process of creating an icon. He needed to achieve the state of *Grace* ("…shine wisely…") and, at the same time, to maintain a certain mental state in whatever concerns the reason and the feeling. All those were necessary to create the possibility for the iconographer (being pure), to create icons with "good pleasure", meaning not hard labor as other toils done by men.

The prayer continued with the specification of the place the icon was worshiped ("…adornment of Your Holy Church…"), its functionality and utility ("…to let go of the sins of those who shyly and with good devotion will kiss and worship them…") and ended with clarifying once again the functionality of the sacred art object created by the iconographer.

Starting as a prayer of praise, the remaining part had the characteristics of a request. This character soon becomes more obvious ("…get him rid of Satan and grow him into Your Commandments…"). In the same part of the prayer we notice that slipping into heresy, deviating from the dogma and tradition, was considered a likely temptation and a devilish impulse. As usual the possibility for such a digression was never considered a consequence of either ignorance or carelessness.

At the end of this prayer, there was a solicitation, again, for Virgin Mary, Luke the Evangelist, and all the saints (as in the foreword of Cennini's handbook) to intermediate the connection between the painter (the creator of sacred art) and the Divinity. The similarity with the position of the western painters definitely shows not only a common position, at least about this issue, in the entire European area, but also a common origin and influences from the old sources as we already mentioned.

The next step during the service we are describing was an important and well-known prayer of solicitation (*ectenie*[9]) and a final request for God to sanction and effect the consecration (bestowing the *Grace* upon the apprentice).

The conclusion of the ceremony was done by *otpust*, the traditional ending of a religious service. In it permission was granted to leave the location where the ceremony took place. This was a small *otpust* in which, as opposed to the great one, the day the service took place was not important. This situation can be considered as an indication that such a consecration was not bound to take place on a certain day or holiday. What is sure is that it could not be scheduled on days of Christian mourning (e.g. Good Friday, which precedes the Easter holiday by two days).

The inclusion of the *otpust* in the ceremony can be considered an indication of the importance given to the consecration of the apprentice and also of the necessity for it to be attended by as many believers as possible who would join their prayers to the priest's. As in any sum of vectors, the resultant was considered to be commensurate with their number. The participation in the ceremony was required from the beginning to the end. The end was marked by the officiating priest granting permission to leave the site of the service. Another sign of the importance of this ceremony was the fact that the ceremony did have as subject only one person.

[9] A prayer in a Christian service in which participants at the service respond to lines spoken by the priest.

We already noted that, as in other more or less similar consecrations, the apprentice (or the journeyman, in the next stage of the training) benefited from the ceremony. He acquired some qualities considered necessary for an activity dedicated to serving the Divinity. The difference compared to similar ceremonies dedicated to future servants of God (e.g. priests) is that, in the case we study, the apprentice or journeyman could not transmit the effects of the consecration to a third party. Starting after consecration, it was considered that the newly consecrated one had a working connection with the Transcendental and the capability to create functional icons. Both qualities were considered as absolutely necessary for a creator of sacred art. After the consecration, the iconographer could transmit only the technico-artistic "secrets" of his craft and he was obliged to do so when he, in turn, took an apprentice.

The same character of consecration marked the ceremony through which the apprentice, at the end of the apprenticeship, was elevated to the rank of journeyman, the intermediate rank between apprentice and iconographer. While there are some indications in the forewords of some Romanian and Athonite *hermeneias* about the consecration at the beginning of the apprenticeship, details about a similar event at the end are offered only by Costin Petrecu [5]. It is mentioned therein that this ceremony was pretty much a repeat of the previous one, marking an important crossroad in the training and activity of a future iconographer.

Why another consecration was considered necessary at this stage, how different was it from the first one—if at all—and if there were new effects expected from it is difficult to say with the amount of information available. Still, we have to point out that following the service at the end of the apprenticeship, the painter was allowed to contract himself and execute—according to his own conception and experience—the painting or the restoration of a church as well as that of fixed or portable icons. When completing the work he was not required anymore to take into account the opinion of a master or to solicit his endorsement in order to contract and execute the order. Additionally, he was not subject to physical punishment by his master in case he made a mistake and was also spared chores in the workshop which were not actually part of the activity of an iconographer.

11.3 The Journeymen

There is not too much information concerning how the promotion of the apprentice to journeyman was done in the religious painting workshops. The details we found in the forewords of the *hermeneias* and in other sources are too scanty for us to form a clear picture about this. The fact that in the manuscripts most of the information concerns the apprentice's consecration can be interpreted as implying that prior to the middle of the nineteenth century it was believed that a consecration done at the beginning of the apprenticeship was sufficient for all the future activity. The few existing testimonies do not seem to offer any argument in favor of a new consecration. Concerning the psychological auto-conditioning, done continuously, it was considered not only necessary but also sufficient for maintaining the state of *Grace* throughout the entire activity.

After Costin Petrescu [4], C. Sandulescu-Verna partially and in passing tackled the topic we address herein in a periodical, more than seven decades ago [21]. He practically discussed only the information offered by the former. Several decades later, in the preface of one of his own compilations [22], he came back with information found in "several Romanian and Byzantine *hermeneias*".

From the statements of the two iconographers mentioned (the latter was a man of the cloth too) it appears that the ceremony was taking place on Easter Day, after the noon service. This information has to be interpreted as saying that in the second half of the nineteenth century (period the two sets of data are coming from), in the area of the historical Romanian province of Muntenia, it was on this great religious holiday that the consecration ceremony of the journeymen took place. As an additional proof for the importance given to iconographers by the Church of the East, we have to mention that, during early Christianity, only certain religious services were officiated which were considered very important (e.g. christening of *catechumens*[10] and of the "called" ones). By scheduling the date of the ceremony on this day, a new dimension is added to the importance the Church was giving it. When and why the ceremony was limited to taking place on Easter Day? Did this happen closer to the present? Why did it happen? We cannot yet answer with the data we have. What we can say so far is that the ceremony of the consecration of the apprentice and journeymen iconographers, at least for a period of time and in a certain area, was officiated the same day as the Resurrection, a day when, since ancient times, the renewal of the entire vegetal life—a phenomenon considered a miracle—was celebrated.

Ten years later Costin Petrescu wrote again [5] about the date these consecrations were taking place. This time he says that the Romanian iconographers were celebrating together the "big religious holidays" and that it was on these occasions that the consecration of the journeymen took place. This statement shows that, at least in his geographical area, the consecrations we are discussing here were not done exclusively on Easter.

Concerning the location of the consecration ceremony, Costin Petrescu and C. Sandulescu-Verna seem to provide different information. The former says that at least in the region where he was born and spent his adolescence "on Easter Day….the masters and journeymen of the whole guild were gathering at the same table …" without clearly specifying if during this brotherly repast consecrations were taking place.

C. Sandulescu-Verna writes about the consecration ceremony for the journeymen as a "religious service", stating specifically that it was taking place in a church. That is why we do not consider that he refers to a certain time period. With only that much detail available from the two sources, the only statement we can make is that, at least in some parts of the Romanian area, this ceremony was officiated in a church.

We go now back to a previous concise remark we made, that there is no mention in the Romanian manuscripts [7–10] or the Athonite *hermeneias* to which we had access [3, 12, 23] about the consecration ceremony of the journeymen. At the present state of the information it cannot be stated with certainty that the date when the

[10] Catechumen–a person who, prior to baptism, receives instruction from a catechist in the principles of the Christian religion.

ceremony was officiated was not moved from the beginning of the apprenticeship to the end of it. This could be supported by the fact we mentioned previously that, as said in the prefaces of the *hermeneias*, only after a first stage in the apprenticeship was the apprentice granted access to all the "secrets" of the craft (canons, iconographic and technical recommendations). The hypothesis that a particular "prayer" (service) was done not only at the beginning but at the end of the apprenticeship period cannot be totally excluded. Maybe this is the reason that the dimensions of the consecration ceremony started to decrease over time.

Sometimes there are differences in the information available even when the sources are apparently related. Sandulescu-Verna's compilation [22] was probably influenced by the content of the foreword of a Romanian manuscript [8] when he discusses aspects in the ceremony of consecration. However, differently from this manuscript, he attributes the entire ceremony to the consecration of the journeyman not the apprentice's, as the *hermeneia* does.

In describing the progress of the consecration ceremony of the journeyman, Costin Petrescu in [5] provides a detail which cannot be found either in the forewords of the *hermeneias* or anywhere else. The author tells us that, on this occasion, the future journeyman was bringing to the ceremony a "diploma project", an icon (subject: the Resurrection) painted by him and a "kerchief full of eggs", painted not dyed in the well-known popular technique. The connection between the subject of the icon, the old custom to offer eggs on Easter Day, the consecration of the journeyman and this important religious holiday is self-explanatory and touching.

The emphasis on this major event in the life of the painter was enhanced by the symbolic motifs with which the eggs were decorated, mentioned briefly in [5]. There were small crosses (the symbol of Christianity), angels (the pure spirits), lambs (the symbol of sacrifice), the rooster who sang three times (according to Christian tradition, to mark how Peter the Apostle denied Jesus, fulfilling an old prophecy), St. Veronica's kerchief (the one miraculously imprinted, not painted by hand), the ladder, the spear, and the sponge infused with vinegar of the Crucifixion, the scissors used by the Roman soldiers to share the cloth covering Jesus, etc. As we can easily see, the motifs are Christian, belong to historical events from the New Testament, are all connected to the Easter holiday and most have no echoes from faiths older than Christianity. It is known that dyeing and decorating eggs is older than that.

After the entire guild had assessed the ability of the journeyman by looking at his icon and the painted eggs, and after the officiating priest "had given his blessing" [5], he said a prayer which we recognize as being the same as the one from the service for the apprentice's consecration. It was listened to by the entire audience which was kneeling. This feature shows the importance given to the prayer in general but also, in particular, the importance of the moment when all those present were soliciting the descent of the Divine Grace unto the journeyman.

By subsequently kissing the priest's hand (gratitude for intermediating the descent of *Grace* and the Heavenly blessing) and the abbot's (humility and obedience towards the first among peers within *Grace* and guild), as well as the cheeks of everybody present (humility, joy, fraternity), the one freshly consecrated and admitted to the circle of initiates was showing his feelings on the occasion of the first independent steps in the world of sacred art. Concerning the homage brought to the abbot, we have to reiterate that he had a certain status within the iconographers' guild. A document hand-issued by the prince Nicolae Mavrocordat in 1786 was giving to a painter (Ioniţa) the "right of judging over the apprentices and journeymen from the painters' guild"[22].

The next step in the consecration ceremony was the hugging of the new journeyman by the abbot, as a welcome in the painters' circle. At this point the abbot was giving the apprentice a gift composed of a set of underwear, a suit, a pair of footwear, and an overcoat lined with fox fur. It was the first gift to the newcomer from the first among peers. This gift may have had other meanings too, but the main one, of human solidarity with a youth started on the life's journey, is the most obvious one.

After having presented the ceremony geared towards obtaining the *Grace*, it is necessary to examine how, through psychological auto-conditioning sustained during his entire activity, the iconographer was able to maintain that state, obtained through consecration. Both the Church and the painter considered this state as a compulsory condition if the icons realized by him were to be functional and have all the required qualities.

11.4 Maintaining the *Grace*

At the beginnings of religious painting the vast majority of the iconographers were monks. The auto-conditioning was easier to realize in such a case. The rigors imposed on the painters were less than those imposed on monks. Many among the painters who were monks too, but also some from those having started as priests, practiced *Hesychasm*[11] in the context of auto-conditioning [2]. This practice has its roots and spirituality in the Byzantine *Hesychast* tradition. The situation was the same over the entire Orthodox East. Some well-informed researchers remarked, for instance, that the renowned Russian painter Rublev, by the way he perceived the events represented in some of his works, showed clearly the influence of the Russian *Hesychastic* center of his time [24].

About the way the life of icon painters was supposed to be lead and about what we called psychological auto-conditioning, even fewer details are available than there are about the previous topics discussed herein. Surely, some of the norms concerning it were considered secrets which could not be disclosed outside a closed circle of initiates. Others were considered common and compulsory for the men of the cloth and for the faithful alike (humility, fasting, charity, some prayers specific to some parts of the day, etc.) and that is why they did not appear in the recommendations concerning the activity of iconographers. Finally, there were others, such as reading, while the painting was being done (of some excerpts from religious

[11] *Hesychasm*—living a solitary life (origin, adj: hermit) focused on the practice of inner praying. It targets acquiring an inner focus and blocking of the physical senses.

books: the Bible, the Lives of Saints, the writings of the Church's Founders), which could be learned while working in a team and which, although publicly known, still had secret meanings and effects. Concerning the readings from religious texts, such a text, read by an apprentice [3] while an area of the church was painted, had a close and direct relationship with the subject being painted [19]. By the aura of incantation and by the prompting to devotion, the monotonous reading, through its meanings and effects, was conducive to creating and maintaining the ambiance absolutely necessary for the creation of a sacred art object, be it a mural, a portable or a fixed icon.

As opposed to the iconographic and technical recommendations which were transmitted through the master-apprentice connection (orally and by practical demonstration), those concerning psychological conditioning and maintaining the state of *Grace* were probably transmitted mostly through the confessor-iconographer connection, the participation of the master being sensibly reduced in this matter.

Both the distant common origins and models and the spiritual unity of the Orthodox East appear clearly where this matter is concerned. For instance, both the Athonite *hermeneia* of Dionysius of Fourna and documents of Russian Synods recommended to iconographers to prepare continuously, by prayer, meditation, devotion, and moderate lifestyle. The Church and the religious painters considered that only continuous auto-conditioning allowed maintaining the state of *Grace* and the contact with the world beyond, in order to represent it and convey its message.

Without being subject to exactly the same rigors, the entire Christian West thought and acted similarly during a certain time. Michelangelo's words illustrate this situation the best: a creator of a "...good painting..." has to "approach God, become one with Him" and "his life has to be holy and immaculate as much as possible, and his spirit be the leader of his thought".

On the same topic of the psychological auto-conditioning of the Southeastern European iconographers, tradition required from them deep beliefs ("fully knowledgeable and adept in the Christian principles"), "growing within your Commendments", "devoted and God fearing", with a Christian life (obedient to some rigors) and with an "elevated living", leading a "harsh, ascetic life", meaning a life governed by the rigors mentioned and with uninterrupted mental contact with a world to which he had dedicated his life and activity [2]. Kindness and tolerance, compassion and Christian solidarity with all men, absence of frivolity, envy and aggressiveness were part of a good Christian's profile too and, implicitly of the iconographer's.

The same forewords in the *hermeneias* we already referred to insisted on a quality the iconographer needed to have, humility. In connection with this, it is very likely that among the motives an Orthodox painter had for (usually) not signing his work, humility was one preventing him from garnering renown for himself as a result of his activity, another being his belief that he was only a means through which the Divine will was carried out. In this area, Costin Petrescu, a third generation icon painter provides a touching piece of information. He says that the painter "was content just to whisper his given name, without any concern but to be mentioned by the priest on holidays and at the Hour of the Holy Mystery". That means, without actually signing the icon.

The closer we come to the present, the higher the number of signed icons becomes. It would be interesting maybe to study whether this upward trend is matched by the descending pattern of the curve representing the frequency and intensity of the auto-conditioning of iconographers and, implicitly, a decline in religious painting.

The situation concerning the signing of icons is similar over the entire Orthodox East. One exception has to be noted. In the Greek area, when the iconographers did sign their names, they would have preceded it by "through the hand of" [22], the same expression they used when they referred to their own activity. It may be of interest to mention that such an inscription ("through the hand of all-humble painters") was found in the Romanian area too, in the main church at Monastery Horezu [25]. Similarly, in the Russian area humility and everything ensuing from it were considered a must for religious painters [25].

Concerning the fasting ("that of food" but also "that of sins"), accompanied by prayers and meditation, as a help in maintaining the state of *Grace*, practicing it was highly appreciated in the world of the iconographers of Southeastern Europe. A testimony to this effect can be found in [26]. In it Mihail Mihalcu describes an interview with a very old iconographer who described how he healed himself of a terrible and unforgiving pulmonary affliction by total fasting ("black fast") for forty days. His belief in the miraculous effect of restraining bodily appetites, the stamp of the traditional iconographer in that area, was impressive and touching. Additionally it appeared to open a window toward a mysterious world, with energies still unknown and unused by the present day man, a world full of fabulous miracles.

It is necessary to recall very briefly other means used in the process of auto-conditioning of the religious painters. Some of the means were: lighting a certain number of candles or votive lights (in church, studio or home) on the occasion of important religious holidays or commemorations, before beginning a big work, a series of icons with particular destination or, periodically, throughout their entire activity. Other means were: a certain number of genuflections daily, completion of some unpaid works in some churches ("founders by work") or pilgrimage to places of importance for the Church or to an icon renowned for accomplishing miracles. Along these lines we have to notice the number of candles, fast days and genuflections were either multiples of three or of forty. These numbers were probably relaying ancient echoes from before Christianity.

Finally, we need to mention that in the training of an apprentice iconographer a particular attention was to be given not only to achieving and maintaining a spiritual purity ("cleanliness") but to the body cleanliness as well. In the effort to attend to the latter some definite rules needed to be known and observed in general in the monastic life as well as during the apprenticeship spent in the workshop and household of an established iconographer.

Some conclusions may be formulated concerning the training and the lifestyle of the religious painter of times past. The Church and the iconographer were looking in the same way at the role, the realization and the functionality of icons and the qualities necessary in a creator of sacred art, and at the way those could be attained and maintained too. Without possessing the *Grace*, obtained through a consecration and maintained by psychological auto-conditioning, it was believed that it is not possible to create a functional icon.

In the above mentioned position of the Church and of the iconographer towards icons and their creators, in time, some changes occurred. Some of these are acknowledged, some are not yet, but they became more and more noticeable starting with the second half of the last century. It is certain that an important contribution to this accentuation was from some changes in the social life, industrialization, development in sciences in general, in economic life, and also from some political utopias which led to devastating social seisms. We can ultimately say that the entire society's conception about world and life has changed [27].

References

1. Mihalcu M, Leonida MD (1981) Les erminies—les "livres" de peinture roumains—témoignages écrits uniques pour l'histoire de la science. The 16th Int. Congress of History of Science (Ext. Abstr.). Bucharest (Aug.26-Sept.2)
2. Mihalcu M, Leonida MD (2009) Din tainele iconarilor romani de altadata. Ed. Eikon, Cluj-Napoca
3. Didron P, Durand AN (1845) Manuel d'iconographie chretienne grecque et latine. Imprimerie Royale, Paris
4. Petrescu C (19 April 1923) Cum zugraveau parintii nostri. Conference at the Romanian Atheneum, Bucharest
5. Petrescu C (1932) O consacrare artistica in ziua de Pasti. Universul (Bucuresti), XLIX, May 1, 20: 9
6. Mihalcu M (1984) Romanian medieval values. Ed. Sport-Turism, Bucharest
7. Romanian manuscript 1555 State Archives, Bucharest
8. Romanian manuscript 1795 Library of the Romanian Academy, Bucharest
9. Romanian manuscript 1283 Library of the Romanian Academy, Bucharest
10. Romanian manuscript 5769, Library of the Romanian Academy, Bucharest
11. Cennini CA (1960) The craftsman's handbook, translated by Daniel Thompson. Dover Publ. Inc., New York
12. Dionysius of Fourna (1974) The painter's manual. Translated by Paul Hetherington, St. Vladimir Seminary Press, Yonkers
13. Voinescu T (1978) Radu Zugravu. Ed. Meridiane, Bucharest
14. Mihalcu M (1995) The patterns used by medieval Romanian painters. Etudes et Documents Balcaniques et Mediterraneens (Paris), Recueil. 18, pp. 23–28
15. Grecu V (1936) Carti de pictura bisericeasca bizantina. Ed. Glasul Bucovinei, Cernauti
16. Romanian manuscript 4206, Library of the Romanian Academy, Bucharest
17. Romanian manuscript 2501, Library of the Romanian Academy, Bucharest
18. Romanian manuscript 1808 Library of the Romanian Academy, Bucharest
19. Mitropolia Olteniei (1967) Craiova, XIX, 5–6: 408–429
20. Duchesne E (1920) Le Stoglav ou les cents chapitres: Recueil de Decisions de l'Assemblee Ecclesiastique de Moscou. Librairie Ancienne Honore Champion, Paris
21. Sandulescu-Verna C (1937) Old painter. The Angel (Buzau). Ingerul IX:1–2, 34–38
22. Sandulescu-Verna C (1979) Erminia picturii bizantine. Ed. Mitropoliei Banatului, Timisoara
23. Schaffer G (1855) Das Handbuch der Malerei vom Berge Athos. Druch und Verlag der Fr. Lintzschen Buchhandlung, Trier
24. Ouspensky L, Lossky V (1989) The meaning of icons. St. Vladimir's Seminary Press, Yonkers
25. Benz E (1971) Geist und leben der Ostkirche. Forum Slavicum, vol. 30, 2nd edn. Wilhelm Fink Verlag, Munich
26. Mihalcu M (2002) Relatia biserica-icoana-iconar. Consacrarea pictorilor bisericesti. Etudes Roumaines et Aroumaines, Paris-Bucharest (VI, 45)
27. Iorga N (1905) Inscriptii din bisericile Romaniei, vol. 1. Minerva, Bucharest

Biographical Note

Mihaela D. Leonida is a Professor of Chemistry at Fairleigh Dickinson University, Metropolitan Campus, in Teaneck, NJ (mleonida@fdu.edu). She received a M.S.and a Ph.D. in Chemical Engineering from Polytechnic University in Bucharest, Romania, and a second Ph.D. in Chemistry from Wesleyan University in Middletown, CT. While teaching, she mentored research with over sixty students.

Her scientific interests are broad, from "wiring" enzymes to bionanomaterials with antimicrobial and anti-proliferation activity, to detecting art forgeries, to materials and techniques used by artisans of times past. She spent a sabbatical leave doing research and lecturing at l'Universite de Quebec a Montreal and at McGill University and was invited to lecture and collaborate on an interdisciplinary project at the Ecole Polytechnique de l'Universite de Nantes. In 2010 she spent one semester as a Fulbright Scholar lecturing (about materials in art and modified enzymes) and doing research at the Polytechnic University in Bucharest. She published four books, two textbooks, and several chapters in books. Dr. Leonida published over sixty papers in scientific journals (among which: Journal of Organic Chemistry, Journal of Analytical and Bioanalytical Chemistry, Nonlinear Optics and Quantum Optics, Internationsl Journal of Nano&Biomaterials, Phytomedicines, Études et Documents Balkaniques et Mediterranéens, International Journal of Bio-Chromatography, Bioorganic & Medicinal Chemistry Letters, Tetrahedron Letters, Revue Roumaine de Chimie, Review of Museums and Historical Monuments, Current Nanoscience) and made over eighty presentations at scientific conferences. She gave over thirty invited lectures at different venues in USA, Canada, France, and Romania.

Dr. Leonida's strong interest in the materials and techniques used by folk artisans, especially iconographers, resulted in two books, several presentations and invited lectures, and lifelong connections and collaborations with researchers from different fields. Most of her interests, work and lectures are focused on interdisciplinary topics. She is actively involved in the American Chemical Society, at the local and national level.

M. D. Leonida, *The Materials and Craft of Early Iconographers*,
SpringerBriefs in Materials, DOI 10.1007/978-3-319-04828-4, © The Author(s) 2014

Index

M. D. Leonida, *The Materials and Craft of Early Iconographers*,
SpringerBriefs in Materials, DOI 10.1007/978-3-319-04828-4, © The Author(s) 2014

MIX
Papier aus verantwortungsvollen Quellen
Paper from responsible sources
FSC® C105338

FSC
www.fsc.org

If you have any concerns about our products,
you can contact us on
ProductSafety@springernature.com

In case Publisher is established outside the EU,
the EU authorized representative is:
Springer Nature Customer Service Center GmbH
Europaplatz 3, 69115 Heidelberg, Germany

Printed by Libri Plureos GmbH
in Hamburg, Germany